Dynamical Systems Method for Solving Operator Equations

This is volume 208 in
MATHEMATICS IN SCIENCE AND ENGINEERING
Edited by C.K. Chui, *Stanford University*

A list of recent titles in this series appears at the end of this volume.

Dynamical Systems Method for Solving Operator Equations

Alexander G. Ramm
DEPARTMENT OF MATHEMATICS
KANSAS STATE UNIVERSITY
USA

ELSEVIER
Amsterdam – Boston – Heidelberg – London – New York – Oxford
Paris – San Diego – San Francisco – Singapore – Sydney – Tokyo

Elsevier
Radarweg 29, PO Box 211, 1000 AE Amsterdam, The Netherlands
The Boulevard, Langford Lane, Kidlington, Oxford OX5 1GB, UK

First edition 2007

Library of Congress Cataloging-in-Publication Data
A catalog record for this book is available from the Library of Congress

British Library Cataloguing in Publication Data
A catalogue record for this book is available from the British Library

ISBN-13: 978-0-444-52795-0
ISBN-10: 0-444-52795-8
ISSN: 0076-5392

Transferred to digital printing in 2007.

07 08 09 10 11 10 9 8 7 6 5 4 3 2 1

Preface

In this monograph a general method for solving operator equations, especially nonlinear and ill-posed, is developed. The method is called the dynamical systems method (DSM). Suppose one wants to solve an operator equation:

$$F(u) = 0, \tag{1}$$

where F is a nonlinear or linear map in a Hilbert or Banach space. We assume that equation (1) is solvable, possibly non-uniquely. The DSM for solving equation (1) consists of finding a map Φ such that the Cauchy problem

$$\dot{u} = \Phi(t, u), \quad u(0) = u_0; \quad \dot{u} = \frac{du}{dt}, \tag{2}$$

has a unique global solution, i.e., solution $u(t)$ defined for all $t \geq 0$, there exists
$u(\infty) = \lim_{t\to\infty} u(t)$, and $F(u(\infty)) = 0$:

$$\exists! u \quad \forall t \geq 0; \quad \exists u(\infty); \quad F(u(\infty)) = 0. \tag{3}$$

If (3) holds, we say that DSM is justified for equation (1). Thus the dynamical system in this book is a synonym to an evolution problem (2). This explains the name DSM. The choice of the initial data $u(0)$ will be discussed for various classes of equations (1). It turns out that for many classes of equations (1) the initial approximation u_0 can be chosen arbitrarily, and, nevertheless, (3) holds, while for some problems the choice of u_0, for which (3) can be established, is restricted to some neighborhood of a solution to equation (1).

We describe various choices of Φ in (2) for which it is possible to justify (3). It turns out that the scope of DSM is very wide. To describe it, let us introduce some notions. Let us call problem (1) *well-posed* if

$$\sup_{u \in B(u_0, R)} ||[F'(u)]^{-1}|| \leq m(R), \tag{4}$$

where $B(u_0, R) = \{u : ||u - u_0|| \leq R\}$, $F'(u)$ is the Fréchet derivative (F-derivative) of the operator-function F at the point u, and the constant $m(R) > 0$ may grow arbitrarily as R grows. If (4) fails, we call problem (1) *ill-posed* . If problem (1) is ill-posed, we write it often as $F(u) = f$ and assume that noisy data f_δ are given in place of f, $||f_\delta - f|| \leq \delta$. Although the equation $F(u) = f$ is solvable, the equation $F(u) = f_\delta$ may have no solutions.

The problem is:

Given $\{\delta, f_\delta, F\}$, find a stable approximation u_δ to a solution u of the equation $F(u) = f$, i.e., find u_δ such that

$$\lim_{\delta \to 0} ||u_\delta - u|| = 0. \tag{5}$$

Unless otherwise stated, we assume that

$$\sup_{u \in B(u_0, R)} ||F^{(j)}(u)|| \leq M_j(R), \quad 0 \leq j \leq 2, \tag{6}$$

where $M_j(R)$ are some constants. In other words, we assume that the nonlinearity is C^2_{loc}, but the rate of its growth, as R grows, is not restricted.

Let us now describe briefly the scope of the DSM.

Any well-posed problem (1) can be solved by a DSM which converges at an exponential rate, i.e.,

$$||u(\infty) - u(t)|| \leq re^{-c_1 t}, \quad ||F(u(t))|| \leq ||F_0||e^{-c_1 t}, \tag{7}$$

where $r > 0$ and $c_1 > 0$ are some constants, and $F_0 := F(u_0)$.

For ill-posed problems, in general, it is not possible to estimate the rate of convergence; depending on the data f this rate can be arbitrarily slow. To estimate the rate of convergence in an ill-posed problem one has to make some additional assumptions about the data f. Remember that by "any" we mean throughout any solvable problem (1).

Any solvable linear equation

$$F(u) = Au - f = 0, \tag{8}$$

where A is a closed, linear, densely defined operator in a Hilbert space H, can be solved stably by a DSM. If noisy data f_δ are given, $||f_\delta - f|| \leq \delta$, then DSM yields a stable solution u_δ for which (5) holds.

We derive stopping rules, i.e., rules for choosing $t(\delta) := t_\delta$, the time at which $u_\delta(t_\delta) = u_\delta$ should be calculated, using f_δ in place of f, in order for (5) to hold.

For linear problems (8) the convergence of a suitable DSM is global with respect to u_0, i.e., DSM converges to the unique minimal-norm solution of y of (8) for any choice of u_0.

Similar results we prove for equations (1) with monotone operators $F : H \to H$. Recall that F is called monotone if

$$(F(u) - F(v), u - v) \geq 0 \quad \forall u, v \in H, \tag{9}$$

where H is a Hilbert space. For hemicontinuous monotone operators the set $\mathcal{N} = \{u : F(u) = 0\}$ is closed and convex, and such sets in a Hilbert space have unique minimal-norm element. A map F is called hemicontinuous if the function $(F(u + \lambda v), w)$ is continuous with respect to $\lambda \in [0, \lambda_0)$ for any $u, v, w \in H$, where $\lambda_0 > 0$ is a number.

DSM is justified for any solvable equation (1) with monotone operators satisfying conditions (6). Note that no restrictions on the growth of $M_j(R)$ as R grows are imposed, so the nonlinearity is C^2_{loc} but may grow arbitrarily fast. For monotone operators we will drop assumption (6) and construct a convergent DSM.

We justify DSM for arbitrary solvable equation (1) in a Hilbert space with C^2_{loc} nonlinearity under a very weak assumption:

$$F'(y) \neq 0, \tag{10}$$

where y is a solution to equation (1).

We justify DSM for operators satisfying the following spectral assumption:

$$||(F'(u) + \varepsilon)^{-1}|| \leq \frac{c}{\varepsilon}, \quad 0 < \varepsilon \leq \varepsilon_0, \quad \forall u \in H, \tag{11}$$

where $\varepsilon_0 > 0$ is an arbitrary small fixed number. Assumption (11) is satisfied, for example, for operators $F'(u)$ whose regular points, i.e., points $z \in \mathbb{C}$ such that $(F'(u) - z)^{-1}$ is a bounded linear operator, fill in the set

$$|z| < \varepsilon_0, \quad |\arg z - \pi| \leq \varphi_0, \tag{12}$$

where $\varphi_0 > 0$ is an arbitrary small fixed number. We also prove the existence of a solution to the equation:

$$F(u) + \varepsilon u = 0, \tag{13}$$

provided that (6) and (11) hold.

We discuss DSM for equations (1) in Banach spaces. In particular, we discuss some singular perturbation problems for equations of the type (13): under what conditions a solution u_ε to equation (13) converges to a solution of equation (1) as $\varepsilon \to 0$.

In Newton-type methods, e.g.,

$$\dot{u} = -[F'(u)]^{-1} F(u), \quad u(0) = u_0, \tag{14}$$

the most difficult and time-consuming part is the inversion of the derivative $F'(u)$.

We propose a DSM method which avoids the inversion of the derivative. For example, for well-posed problem (1) such a method is

$$\begin{aligned} \dot{u} &= -QF(u), \quad u(0) = u_0, \\ \dot{Q} &= -TQ + A^*, \quad Q(0) = Q_0, \end{aligned} \tag{15}$$

where

$$A := F'(u), \quad T = A^*A, \tag{16}$$

A^* is the adjoint to A operator, and u_0 and Q_0 are suitable initial approximations.

We also give a similar DSM scheme for solving ill-posed problem (1).

We justify DSM for some classes of operator equations (1) with unbounded operators, for example, for operators $F(u) = Au + g(u)$ where A is a linear, densely defined, closed operator in a Hilbert space H and g is a nonlinear C^2_{loc} map.

We justify DSM for equations (1) with some nonsmooth operators, e.g., with monotone, hemicontinuous, defined on all of H operators.

We show that the DSM can be used as a theoretical tool for proving conditions sufficient for the surjectivity of a nonlinear map or for this map to be a global homeomorphism.

One of our motivations is to develop a general method for solving operator equations, especially nonlinear and ill-posed. The other motivation is to develop a general approach to constructing convergent iterative processes for solving these equations.

The idea of this approach is straightforward: if the DSM is justified for solving equation (1), i.e., (3) holds, then one considers a discretization of (2), for example:

$$u_{n+1} = u_n + h_n\Phi(t_n, u_n), \quad u_0 = u_0, \quad t_{n+1} = t_n + h_n, \tag{17}$$

and if one can prove convergence of (17) to the solution of (2), then (17) is a convergent iterative process for solving equation (1).

We prove that any solvable linear equation (8) (with bounded or unbounded operator A) can be solved by a convergent iterative process which converges to the unique minimal-norm solution of (8) for any initial approximation u_0.

A similar result we prove for solvable equation (1) with monotone operators.

For general nonlinear equations (1), under suitable assumptions, a convergent iterative process is constructed. The initial approximation in this process does not have to be in a suitable neighborhood of a solution to (1).

We give some numerical examples of applications of the DSM. A detailed discussion of the problem of stable differentiation of noisy functions is given.

New technical tools, that we often usein this book, are some novel differential inequalities.

The first of these deals with the functions satisfying the following inequality:

$$\dot{g} \le -\gamma(t)g(t) + \alpha(t)g^2(t) + \beta(t), \quad t \ge t_0 \ge 0, \tag{18}$$

where g, γ, α, β are nonnegative functions, and γ, α and β are continuous on $[t_0, \infty)$. We assume that there exists a positive function $\mu \in C^1[t_0, \infty)$, such that

$$0 \le \alpha(t) \le \frac{\mu(t)}{2}\left(\gamma(t) - \frac{\dot{\mu}(t)}{\mu(t)}\right), \quad \beta(t) \le \frac{1}{2\mu(t)}\left(\gamma(t) - \frac{\dot{\mu}(t)}{\mu(t)}\right), \tag{19}$$

$$\mu(t_0)g(t_0) < 1, \tag{20}$$

and prove that under the above assumptions, any nonnegative solution $g(t)$ to (18) is defined on $[t_0, \infty)$ and satisfies the following inequality:

$$0 \le g(t) \le \frac{1 - \nu(t)}{\mu(t)} < \frac{1}{\mu(t)}, \tag{21}$$

where

$$\nu(t) = \frac{1}{\frac{1}{1-\mu(t_0)g(t_0)} + \frac{1}{2}\int_{t_0}^t \left(\gamma(s) - \frac{\dot{\mu}(s)}{\mu(s)}\right) ds}. \tag{22}$$

The other inequality, which we use, is an operator version of the Gronwall inequality. Namely, assume that:

$$\dot{Q} = -T(t)Q(t) + G(t), \quad Q(0) = Q_0, \tag{23}$$

where $T(t)$ and $G(t)$ are linear bounded operators on a Hilbert space depending continuously on a parameter $t \in [0, \infty)$. If there exists a continuous positive function $\varepsilon(t)$ on $[0, \infty)$ such that

$$(T(t)h, h) \ge \varepsilon(t)||h||^2 \quad \forall h \in H, \tag{24}$$

then the solution to (23) satisfies the inequality:

$$||Q(t)|| \le e^{-\int_0^t \varepsilon(x)dx}\left[||Q_0|| + \int_0^t ||G(s)||e^{\int_0^s \varepsilon(x)dx} ds\right]. \tag{25}$$

This inequality shows that $Q(t)$ is a bounded linear operator whose norm is bounded uniformly with respect to t if

$$\sup_{t\geq 0}\int_0^t ||G(s)||e^{-\int_s^t \varepsilon(x)dx}ds < \infty. \tag{26}$$

The DSM is shown to be useful as a tool for proving theoretical results, see Chapter 13.

The DSM is used in Chapter 14 for construction of convergent iterative processes for solving operator equation.

In Chapter 15 some numerical problems are discussed, in particular, the problem of stable differentiation of noisy data.

In Chapter 16 various auxiliary material is presented. Together with some known results, available in the literature, some less known results are included: a necessary and sufficient condition for compactness of embedding operators and conditions for the continuity of the solutions to operator equations with respect to a parameter.

The table of contents gives a detailed list of topics discussed in this book.

Contents

Chapter 1

Introduction

1.1 What this book is about

This book is about a general method for solving operator equations

$$F(u) = 0. \tag{1.1.1}$$

Here F is a nonlinear map in a Hilbert space H. Later on we consider maps F in Banach spaces as well. The general method, that we develop in this book and call the dynamical systems method (DSM), consists of finding a nonlinear map $\Phi(t, u)$ such that the Cauchy problem

$$\dot{u} = \Phi(t, u), \quad u(0) = u_0, \tag{1.1.2}$$

has a unique global solution $u(t)$, that is, the solution defined for all $t \geq 0$, this solution has a limit $u(\infty)$:

$$\lim_{t \to \infty} ||u(\infty) - u(t)|| = 0, \tag{1.1.3}$$

and this limit solves equation (1.1.1):

$$F(u(\infty)) = 0. \tag{1.1.4}$$

Let us write these three conditions as

$$\exists! u(t) \quad \forall t \geq 0; \quad \exists u(\infty); \quad F(u(\infty)) = 0. \tag{1.1.5}$$

If (1.1.5) holds for the solution to (1.1.2) then we say that a DSM is justified for solving equation (1.1.1). There may be many choices of $\Phi(t, u)$ for which DSM can be justified. A number of such choices will be given in

Chapter 3 and in other Chapters. It should be emphasized that we do not assume that equation (1.1.1) has a unique solution. Therefore the solution $u(\infty)$ depends on the initial approximation u_0 in (1.1.2). The choice of u_0 in some cases is not arbitrary and in many cases this choice is arbitrary, for example, for problems with linear operators, nonlinear monotone operators, and for a wide class of general nonlinear problems (see Chapters 4, 6, 7-9, 11-12, 14).

The existence and uniqueness of the local solution to problem (1.1.2) is guaranteed, for example, by a Lipschitz condition imposed on Φ:

$$||\Phi(t,u) - \Phi(t,v)|| \le L||u-v||, \quad u,v \in B(u_0,R), \tag{1.1.6}$$

where the constant L does not depend on $t \in [0,\infty)$ and

$$B(u_0,R) = \{u : \quad ||u-u_0|| \le R\}$$

is a ball, centered at the element $u_0 \in H$ and of radius $R > 0$.

1.2 What the DSM (Dynamical Systems Method) is

The DSM for solving equation (1.1.1) consists of finding a map $\Phi(t,u)$ and an initial element u_0 such that conditions (1.1.5) hold for the solution to the evolution problem (1.1.2).

If conditions (1.1.5) hold, then one solves Cauchy problem (1.1.2) and calculates the element $u(\infty)$. This element is a solution to equation (1.1.1). The important question one faces after finding a nonlinearity Φ, for which (1.1.5) hold, is the following one: how does one solve Cauchy problem (1.1.2) numerically? This question has been studied much in the literature. If one uses a projection method, i.e., looks for the solution of the form:

$$u(t) = \sum_{j=1}^{J} u_j(t) f_j, \tag{1.2.1}$$

where $\{f_j\}$ is an orthonormal basis of H, and $J > 1$ is an integer, then problem (1.1.2) reduces to a Cauchy problem for a system of J nonlinear ordinary differential equations for the scalar functions $u_j(t)$, $1 \le j \le J$, if the right-hand side of (1.1.2) is projected onto the J-dimensional subspace spanned by $\{f_j\}_{1\le j\le J}$. This system is:

$$\dot{u}_j = \left(\Phi(\sum_{m=1}^{J} u_m(t) f_m, t), f_j \right), \quad 1 \le j \le J, \tag{1.2.2}$$

$$u_j(0) = (u_0, f_j), \quad 1 \le j \le J. \tag{1.2.3}$$

Numerical solution of the Cauchy problem for systems of ordinary differential equations has been much studied in the literature.

In this book the main emphasis is on the possible choices of Φ which imply properties (1.1.5).

1.3 The scope of the DSM

One of our aims is to show that DSM is applicable to a very wide variety of problems.

Specifically, we prove in this book that the DSM is applicable to the following classes of problems:

1. *Any well-posed solvable problem (1.1.1) can be solved by DSM.*

   By a *well-posed problem* (1.1.1) we mean the problem with the operator F satisfying the following assumptions:

   $$\sup_{u \in B(u_0,R)} ||[F'(u)]^{-1}|| \leq m(R), \tag{1.3.1}$$

   and

   $$\sup_{u \in B(u_0,R)} ||F^{(j)}(u)|| \leq M_j(R), \quad 0 \leq j \leq 2, \tag{1.3.2}$$

   where $F^{(j)}(u)$ is the j-th Fréchet derivative of F.

   If assumption (1.3.1) does not hold, but (1.3.2) holds, we call problem (1.1.1) *ill-posed.* This terminology is not quite standard. The standard notion of an ill-posed problem is given in Section 2.1.

   We prove that for any solvable well-posed problem not only the DSM can be justified, i.e., Φ can be found such that for problem (1.1.2) conclusions (1.1.5) hold, but, in addition, the convergence of $u(t)$ to $u(\infty)$ is exponentially fast:

   $$||u(t) - u(\infty)|| \leq re^{-c_1 t}, \tag{1.3.3}$$

   where $r > 0$ and $c_1 > 0$ are constants, and

   $$||F(u(t))|| \leq ||F_0||e^{-c_1 t}, \quad F_0 := F(u_0). \tag{1.3.4}$$

2. *Any solvable linear ill-posed problem can be solved by DSM.*

   A linear problem (1.1.1) is a problem

   $$Au = f, \tag{1.3.5}$$

where A is a linear operator. This operator we always assume closed and densely defined. Its null space is denoted

$$\mathcal{N}(A) := \{u : Au = 0\},$$

and its domain is denoted $D(A)$, and its range is denoted $R(A)$.

For a linear ill-posed problems a DSM can be justified and Φ can be found such that convergence (1.1.3) holds for *any* initial approximation u_0 in (1.1.2) and $u(\infty)$ is the unique minimal-norm solutions y to (1.1.5). However, in general, one cannot estimate the rate of convergence: it can be as slow as one wishes if f is chosen suitably. To obtain a rate of convergence for an ill-posed problem one has to make additional assumptions on f. One can give a stable approximation to the minimal-norm solution y to problem (1.1.5) using DSM. This stable approximation u_δ should be found from the noisy data $\{f_\delta, \delta\}$, where f_δ, the noisy data, is an arbitrary element satisfying the inequality

$$||f_\delta - f|| \leq \delta, \tag{1.3.6}$$

and $\delta > 0$ is a small number. The stable approximation is the approximation for which one has

$$\lim_{\delta \to 0} ||u_\delta - y|| = 0. \tag{1.3.7}$$

When one uses a DSM for stable solution of an ill-posed problem (1.1.5), or, more generally, of a nonlinear problem

$$F(u) = f, \tag{1.3.8}$$

then one solves the Cauchy problem (1.1.2), where Φ depends on the noisy data f_δ, and one stops the calculation of the corresponding solution $u_\delta(t)$ at a time t_δ, which is called the stopping time. The stopping time should be chosen so that

$$\lim_{\delta \to 0} ||u_\delta(t_\delta) - u(\infty)|| = 0, \tag{1.3.9}$$

where $u(\infty)$ is the limiting value of the solution $u(t)$ to problem (1.1.2) corresponding to the exact data f. In Chapters 4, 6, and 7 we give some methods for choosing the stopping times for solving ill-posed problems.

3. *Any solvable ill-posed problem (1.3.8) with a monotone operator F, satisfying (1.3.2), can be solved stably by a DSM.*

   If the operator F in problem (1.3.8) is monotone, i.e.,

$$(F(u) - F(v), u - v) \geq 0, \tag{1.3.10}$$

   and assumption (1.3.2) holds, then one can find such Φ that (1.1.5) holds.

   Moreover, *convergence (1.1.3) holds for any initial approximation u_0 in (1.1.2), and $u(\infty)$ is the unique minimal-norm solution y to (1.3.8).*

   If noisy data f_δ, $||f_\delta - f|| \leq \delta$, are given in place of the exact data f, then one integrates the Cauchy problem (1.1.2) with Φ corresponding to f_δ and calculates the corresponding solution $u_\delta(t)$ at a suitably chosen stopping time t_δ.

   If $u_\delta := u_\delta(t_\delta)$ then

$$\lim_{\delta \to 0} ||u_\delta - y|| = 0. \tag{1.3.11}$$

   Some methods for finding the stopping time are discussed in Chapter 6.

4. *Any solvable ill-posed problem (1.3.8), such that*

$$F(y) = f, \quad F'(y) \neq 0,$$

   and (1.3.2) holds, can be solved stably by a DSM.

5. *Any solvable ill-posed problem (1.3.8) with a monotone, hemicontinuous, defined on all of H operator F can be solved stably by a DSM.*

   For such operators assumption (1.3.2) is dropped. One can choose such Φ that convergence (1.1.3) holds for any initial approximation u_0 in (1.1.2) and $u(\infty) = y$, where y is the unique minimal-norm solution to (1.3.8).

6. *If $F = L + g$, where L is a linear, closed, densely defined operator, g is a nonlinear operator satisfying (1.3.2), and equation (1.3.8) is solvable, then it can be solved by a DSM provided that L^{-1} exists, is bounded, and*

$$\sup_{u \in B(u_0, R)} ||(I + L^{-1}g'(u))^{-1}|| \leq m(R). \tag{1.3.12}$$

   Thus DSM can be used for some equations (1.1.1) with unbounded operators F.

7. *DSM can be used for proving theoretical results.*

   For example:

   A map $F : H \to H$ is surjective if (1.3.1)-(1.3.2) hold and

$$\sup_{R>0} \frac{R}{m(R)} = \infty. \tag{1.3.13}$$

   A map $F : H \to H$ is a global homeomorphism of H onto H if (1.3.2) holds and

$$||[F'(u)]^{-1}|| \leq h(||u||), \tag{1.3.14}$$

   where $h(s) > 0$ is a continuous function on $[0, \infty)$ such that

$$\int_0^\infty h^{-1}(s)ds = \infty. \tag{1.3.15}$$

8. *DSM can be used for solving nonlinear well-posed and ill-posed problems (1.1.1) without inverting the derivative $F'(u)$.*

   For example, if assumptions (1.3.1)-(1.3.2) hold and problem (1.1.1) is solvable, then the DSM

$$\begin{cases} \dot{u} & = & -QF(u) \\ \dot{Q} & = & -TQ + A^* \\ u(0) & = & u_0, \quad Q(0) = Q_0, \end{cases} \tag{1.3.16}$$

   converges to a solution of problem (1.1.1) as $t \to \infty$, and (1.1.5) holds. Here Q is an operator,

$$A := F'(u), \quad T := A^*A, \tag{1.3.17}$$

   and A^* is the adjoint to A operator.

   Note that a Newton-type method for solving equation (1.1.1) by a DSM is of the form:

$$u = -[F'(u)]^{-1}F(u), \quad u(0) = u_0. \tag{1.3.18}$$

   This method is applicable to the well-posed problems only, because it requires $F'(u)$ to be boundedly invertible. Its regularized versions are applicable to many ill-posed problems also, as we demonstrate in this book. In practice the numerical inversion of $F'(u)$ is the most difficult and time-consuming part of the solution of equation (1.1.1) by the Newton-type methods. The DSM (1.3.16) avoids completely the inversion of the derivative $F'(u)$. Convergence of this method is proved in Chapter 10, where a DSM scheme, similar to (1.3.16), is constructed for solving ill-posed problems (1.1.1).

9. *DSM can be used for solving equations (1.1.1) in Banach spaces.*

   In particular, if $F : X \to X$ is an operator in a Banach space X and the following spectral assumption holds:

$$||A_\varepsilon^{-1}|| \le \frac{c}{\varepsilon}, \quad 0 < \varepsilon < \varepsilon_0, \tag{1.3.19}$$

   where $c > 0$ is a constant,

$$A_\varepsilon := A + \varepsilon I, \quad \varepsilon = \text{const} > 0, \tag{1.3.20}$$

   and $\varepsilon_0 > 0$ is an arbitrary small fixed number, then the DSM can be used for solving the equation

$$F(u) + \varepsilon u = 0. \tag{1.3.21}$$

10. *DSM can be used for construction of convergent iterative schemes for solving equation (1.1.1).*

    The general idea is simple. Suppose that a DSM is justified for equation (1.1.1). Consider a discretization of (1.1.2)

$$u_{n+1} = u_n + h_n \Phi(t_n, u_n), \quad u_0 = u_0, \quad t_{n+1} = t_n + h_n, \tag{1.3.22}$$

    or some other discretization scheme. Assume that the scheme (1.3.22) converges:

$$\lim_{n\to\infty} u_n = u(\infty). \tag{1.3.23}$$

    Then (1.3.22) is a convergent iterative scheme for solving equation (1.1.1) because $F(u(\infty)) = 0$.

    It is clear now that the DSM has a very wide range of applicability. The author hopes that some numerical schemes for solving operator equations (1.1.1), which are based on the DSM, will be more efficient than some of the currently used numerical methods.

1.4 A discussion of DSM

The reader may ask the following question:

Why would one like to solve problem (1.1.2) in order to solve a simpler looking problem (1.1.1)?

The answer is:

First, one may think that problem (1.1.1) is simpler than problem (1.1.2), but, in fact, this thinking may not be justified. Indeed, if problem (1.1.1) is ill-posed and nonlinear, then there is no general method for solving this problem, while one may try to solve problem (1.1.2) by using a projection method and solving the Cauchy problem (1.2.2)-(1.2.3).

Secondly, there is no clearly defined measure of the notion of the simplicity of problem (1.1.1) as compared with problem (1.1.2). As we have mentioned in Section 1.2, the numerical methods for solving (1.2.2)-(1.2.3) have been studied in the literature extensively (see e.g. [HW]).

The attractive features of the DSM are: its wide applicability; its flexibility: there are many choices of Φ for which one can justify DSM, i.e., prove (1.1.5), and many methods for solving the Cauchy problem (1.1.2); its numerical efficiency: we show some evidences of this efficiency in Chapter 15. In particular, one can solve such classical problems as stable numerical differentiation of noisy data, solving ill-conditioned linear algebraic systems, and other problems, more accurately and efficiently by a DSM than by more traditional methods.

1.5 Motivations

The motivations for the development of the DSM in this book are the following ones.

First, we want to develop a general method for solving linear and, especially, nonlinear operator equations. This method is developed especially, but not exclusively, for solving nonlinear ill-posed problems.

Secondly, we want to develop a general method for constructing convergent iterative methods for solving nonlinear ill-posed problems.

Chapter 2

Ill-posed problems

In this Chapter we discuss various methods for solving ill-posed problems.

2.1 Basic definitions. Examples

Consider an operator equation

$$F(u) = f, \tag{2.1.1}$$

where $F : X \to Y$ is an operator from a Banach space X into a Banach space Y.

Definition 2.1.1. *Problem (2.1.1) is called well-posed (in the sense of J. Hadamard) if F is injective, surjective and has continuous inverse. If the problem is not well-posed, then it is called ill-posed.*

Ill-posed problems are of great interest in applications. Let us give some examples of ill-posed problems which are of interest in applications.

Example 2.1.1. *Solving linear algebraic systems with ill-conditioned matrices.*

Let

$$Au = f, \tag{2.1.2}$$

be a linear algebraic system in $\mathbb{R}^n$, $u, f \in \mathbb{R}^n$, $A = (a_{ij})_{1\le i,j\le n}$ is an ill-conditioned matrix, i.e., the condition number $\kappa(A) = ||A||\,||A^{-1}||$ is large. This definition of the condition number preassumes that A is nonsingular, that is, $\mathcal{N}(A) = \{0\}$. If A is singular, i.e., $\mathcal{N}(A) \ne \{0\}$, then formally $\kappa(A) = \infty$ because $||A^{-1}|| = \infty$. Indeed,

$$||A^{-1}|| = \sup_{f\ne 0} \frac{||A^{-1}f||}{||f||} = \sup_{u=A^{-1}f\ne 0} \frac{1}{\frac{||Au||}{||u||}} = \frac{1}{\inf_{u\ne 0}\frac{||Au||}{||u||}} = \infty, \tag{2.1.3}$$

because $Au = 0$ for some $u \neq 0$ if $\mathcal{N}(A) \neq \{0\}$.

Problem (2.1.2) is practically ill-posed if $\mathcal{N}(A) \neq \{0\}$ but $\kappa(A) \gg 1$, i.e., $\kappa(A)$ is very large. Indeed, in this case small variations Δf of f may cause large variations Δu of the solution u. One has

$$\frac{||\Delta u||}{||u||} = \frac{||A^{-1}\Delta f||}{||A^{-1}f||} \leq \frac{||\Delta f|| ||A^{-1}||}{||f|| ||A||^{-1}} = \kappa(A)\frac{||\Delta f||}{||f||}, \tag{2.1.4}$$

where we have used the inequality $||A^{-1}f|| \geq ||f|| ||A||^{-1}$. If the equality sign is achieved in (2.1.4) then a relative error $\frac{||\Delta f||}{||f||}$ causes $\kappa(A)\frac{||\Delta f||}{||f||}$ relative error $\frac{||\Delta u||}{||u||}$ of the solution. If $\kappa(A) = 10^6$ then the relative error of the solution is quite large.

An example of an ill-conditioned matrix is Hilbert's matrix .

$$h_{ij} = \frac{1}{1+i+j}, \quad 0 \leq i,j \leq n. \tag{2.1.5}$$

Its condition number is of order 10^{13} for $n = 9$. A 2×2 matrix

$$A = \begin{pmatrix} 4.1 & 2.8 \\ 9.7 & 6.6 \end{pmatrix}, \tag{2.1.6}$$

has condition number 2, 249.5. Equation (2.1.2) with A defined by (2.1.6) and $u = \begin{pmatrix} 1 \\ 0 \end{pmatrix}$ is satisfied if $f = \begin{pmatrix} 4.1 \\ 9.7 \end{pmatrix}$. If $f_\delta = \begin{pmatrix} 4.11 \\ 9.70 \end{pmatrix}$ then the corresponding solution $u_\delta = \begin{pmatrix} 0.34 \\ 0.97 \end{pmatrix}$. One can see that a small perturbation of f produces a large perturbation of the solution.

The Hilbert matrix for all $n \geq 0$ is positive-definite, because it is a Gramian of a system of linearly independent functions:

$$h_{ij} = \int_0^1 x^{i+j}\,dx.$$

Example 2.1.2. *Stable summation of the Fourier series and integrals with randomly perturbed coefficients.*

Suppose that

$$f = \sum_{j=1}^{\infty} c_j h_j(x), \tag{2.1.7}$$

where $(h_i, h_j) = \delta_{ij}$, where $\delta_{ij} = \begin{cases} 0, & i \neq j, \\ 1, & i = j, \end{cases}$ and $c_j = (f, h_j)$, where (f, h) is the inner product in a Hilbert space $H = L^2(D)$.

Let us assume that $\{c_{j\delta}\}_{1\leq j<\infty}$ are given, and $\sup_j |c_{j\delta} - c_j| \leq \delta$. The problem is to estimate f given the set $\{\delta, c_{j\delta}\}_{1\leq j<\infty}$.

If f_δ is an estimate of f and $\lim_{\delta\to 0} ||f_\delta - f|| = 0$, then the estimate is called stable. Here $||f|| = (f, f)^{1/2}$.

Methods for calculating a stable estimate of f given noisy data will be discussed later.

Example 2.1.3. *Stable numerical differentiation of noisy data.*

Suppose that $f \in C^2([0,1])$ is not known, but the noisy data $f_\delta \in L^\infty([0,1])$ are given and it is assumed that $||f - f_\delta|| \leq \delta$, where the norm is L^∞-norm.

The problem is:

Given the noisy data $\{\delta, f_\delta\}$, *estimate stably* f'.

We prove that this problem, as stated, does not have a solution. In order to solve this problem one has to have additional information, namely one has to assume an a priori bound

$$||f^{(a)}|| \leq M_a, \quad a > 1, \tag{2.1.8}$$

where $f^{(a)}$ is the derivative of order a. If a is not an integer one defines M_a as follows. Let $a = m + b$, where m is an integer and $0 < b < 1$.

Then

$$||f^{(a)}|| = ||f^{(m)}|| + \sup_{x,s\in[0,1]} \frac{|f^{(m)}(x) - f^{(m)}(s)|}{|x-s|^b}. \tag{2.1.9}$$

One can prove that the data $\{\delta, f_\delta, M_a\}$ with any fixed $a > 1$ allow one to construct a stable approximation of f and to estimate the error of this approximation. For example, this error is $O(\delta^{\frac{1}{2}})$ if $a = 2$, and $O(\delta^{\frac{b}{1+b}})$ if $a = 1 + b,\ 0 < b < 1$.

Usually in the literature the stable approximation of f is understood as an estimate $R_\delta f_\delta$ such that

$$\lim_{\delta\to 0} ||R_\delta f_\delta - f'|| = 0. \tag{2.1.10}$$

The author had introduced ([R35]) a new definition of the stable approximation of f namely, the following one:

Let us call the estimate $R_\delta f_\delta$ of f' stable if:

$$\lim_{\delta\to 0} \sup_{f\in K(\delta,a)} ||R_\delta f_\delta - f'|| = 0, \tag{2.1.11}$$

where

$$K(\delta, a) := \{f : ||f^{(a)}|| \leq M_a, \quad ||f - f_\delta|| \leq \delta\}. \tag{2.1.12}$$

The new definition (2.1.11) - (2.1.12) has an advantage over the standard definition because in the new definition there is no dependence on f (and remember that f is unknown), in contrast with the standard definition, which uses the unknown f.

The estimate $R_\delta f_\delta$ has to be constructed on the basis of the known data $\{\delta, f_\delta, M_a\}$ only. These data may correspond to any f in the set $K(\delta, a)$. Since f is not known, and can be any element from the set (2.1.12), it is more natural to define the stable approximation of f by formula (2.1.11) rather than by formula (2.1.10).

A detailed study of the practically important problem of stable numerical differentiation will be given in Section 15.2.

Example 2.1.4. *Stable solution of Fredholm integral equations of the first kind.*

Consider the equation:

$$Au = f, \quad Au = \int_D A(x,y)u(y)dy, \tag{2.1.13}$$

where $D \subset \mathbb{R}^n$ is a bounded domain and the function $A(x,y) \in L^2(D \times D)$ or

$$\sup_{x \in D} \int_D |A(x,y)|dy \leq M. \tag{2.1.14}$$

If $A \in L^2(D \times D)$ then the operator A in (2.1.13) is compact in $H = L(D)$. If (2.1.14) holds, and

$$\lim_{h \to 0} \sup_{|x-s| \leq h} \int_D |A(x,y) - A(s,y)|dy = 0, \tag{2.1.15}$$

then the operator A in (2.1.13) is compact in $X = C(D)$.

Indeed, if $A \in L^2(D \times D)$, then

$$\int_D \int_D |A(x,y)|^2 dxdy < \infty. \tag{2.1.16}$$

In this case the operator A in (2.1.13) is a Hilbert-Schmidt (HS) operator, which is known to be compact in $H = L^2(D)$.

For convenience of the reader let us prove the following known [KA] result:

Theorem 2.1.1. *Integral operator (2.1.13) is compact as an operator from $L^p(D)$ into $L^q(D)$ if*

$$\int_D \int_D |A(x,y)|^{r'} dxdy \leq M^{r'}, \quad r = \min(p,q), \tag{2.1.17}$$

where $M = const > 0, \quad q' = \frac{q}{q-1}, \quad r' = \frac{r}{r-1}, \quad p, q \geq 1,$ *and*

$$||A||_{L^p \to L^q} := ||A|| \leq M|D|^{\frac{q-r(q-1)}{qr}}, \quad |D| := meas\ D. \tag{2.1.18}$$

Remark 2.1.1. If $p = q = r = r' = 2$, then (2.1.18) yields $||A|| \leq M$, so $measD$ can be infinite in this case, but $||A|| \leq M$.

Proof. If $p \geq r$ then one has

$$|(Au)(x)| \leq \left(\int_D |A(x,y)|^{r'} dy\right)^{\frac{1}{r'}} ||u||_{L^r} \leq \left(\int_D |A(x,y)|^{r'} dy\right)^{\frac{1}{r'}} ||u||_{L^p},$$

where $L^s = L^s(D)$.

Using Hölder's inequality, one gets

$$\left(\int_D |(Au)(x)|^q dx\right)^{\frac{1}{q}} \leq ||u||_{L^p} \left(\int_D dx\right)^{\frac{q-r(q-1)}{qr}} \left(\int_D dx \int_D |A(x,y)|^{r'} dy\right)^{\frac{1}{r'}}. \tag{2.1.19}$$

Thus, estimate (2.1.18) is proved.

To prove the compactness of A, note that estimate (2.1.17) implies that $A(x,y) \in L^2(D \times D)$. Therefore there is a finite - rank kernel $A_m(x,y) = \sum_{j,k=1}^{m} a_j(x) b_k(y)$, which approximates $A(x,y)$ in the $L^{r'}(D \times D)$ with arbitrary accuracy, provided that m is sufficiently large. Therefore

$$\lim_{m \to \infty} ||A - A_m||_{L^p \to L^q} = 0, \tag{2.1.20}$$

by estimate (2.1.18). Since $A_m(x,y)$ is a finite-rank kernel, the corresponding operator A with this kernel is compact. Thus the operator A is compact being the limit of a sequence of compact operators A in the operator norm.

Theorem 2.1.1 is proved. □

It is well known that a linear compact operator in an infinite-dimensional Banach space X cannot have a bounded inverse. Indeed if A is a linear compact operator in X and B is its bounded inverse, then $BA = I$, where I is the identity operator, and $I = BA$ is compact as a product of a compact and bounded operator. But the identity operator is compact only in a finite-dimensional Banach space. Therefore B cannot be bounded if it exists. Consequently, problem (2.1.13) is ill-posed.

Some methods for stable solution of equation (2.1.13), given the noisy data, are developed in this chapter.

Example 2.1.5. *Analytic continuation.*

Let f be an analytic function in a domain D on a complex plane. Assume that f is known on a set $E \subset C$, which has a limit point inside D. Then, by the well-known uniqueness theorem, the function f is uniquely determined everywhere in D. The problem of analytical continuation of f from the set E to D is ill-posed. Indeed, if the noisy data f are given on the set E, such that $\sup_{z\in E} |f_\delta(z) - f(z)| \le \delta$, then $f_\delta(z)$ may not be an analytic function in D, and in this case it may be not defined in D, or, if $f_\delta(z)$ is analytic in D, its values in D can differ very much from the values of f. Consider a simple example:

$$f(z) = e^z, \quad D = \{z : |z| < 1\}, \quad E = \{z : |z| \le a\}.$$

Let $a = 10^{-5}$, $f_\delta(z) = \frac{1}{1-z}$. Then one has

$$\max_{|z|\le a} \left| e^z - \frac{1}{1-z} \right| = \max_{|z|\le a} \left| \sum_{n=2}^{\infty} z^n \left(\frac{1}{n!} - 1 \right) \right| \le \frac{a^2}{2} + \frac{a^3}{1-a} := \delta$$

If $a \le 10^{-5}$ then $\delta < 10^{-10}$. However at $z_0 = 1 - 10^{-5}$ one has

$$|e^{z_0} - \frac{1}{1-z_0}| < 10^5.$$

Example 2.1.6. *The Cauchy problem for elliptic equations.*

Suppose that

$$\Delta u = 0 \text{ in } D \subset \mathbb{R}^n, \quad u|_S = f, \quad u_N|_S = h, \tag{2.1.21}$$

where D is a domain, S is its boundary, and N is the unit outer normal to S.

Finding u from the data $\{f, h\}$ is an ill-posed problem: small perturbation of the data $\{f, h\}$ may lead to the pair $\{f_\delta, h_\delta\}$ which does not correspond to any harmonic function in D. The function h in the data (2.1.21) cannot be chosen arbitrarily. One knows that f alone determines u in (2.1.21) uniquely. Therefore f determines h uniquely as well. The map

$$\Lambda : f \to h$$

is called the Dirichlet-to-Neumann map. This map is injective and its properties are known.

Example 2.1.7. *Minimization problems.*

Let

$$f(u) \geq m > -\infty$$

be a continuous functional in a Banach space X. Consider the problem of finding its global minimum

$$m = \inf_u f(u)$$

and its global minimizer y,

$$f(y) = m.$$

We assume that the global minimizer exists and is unique, and that $m > -\infty$. While the problem of finding global minimum is well-posed, the problem of finding global minimizer is ill-posed. Let us explain these claims. Consider $f_\delta(u) = f(u) + g_\delta(u)$, where $\sup_{u \in X} |g_\delta(u)| \leq \delta$. One has

$$\inf_u [f(u) + g_\delta(u)] \leq \inf_u f(u) + \sup g_\delta(u) \leq m + \delta,$$

and

$$m - \delta \leq \inf_u f(u) - \sup_u |g_\delta| \leq \inf_u [f(u) + g_\delta(u)].$$

Thus

$$m - \delta \leq \inf_u [f(u) + g_\delta(u)] \leq m + \delta, \tag{2.1.22}$$

provided that

$$\sup_u |g_\delta(u)| \leq \delta.$$

This proves that small perturbations of f lead to small perturbations of the global minimum.

The situation with global minimizer is much worse: small perturbation of f can lead to large perturbations of the global minimizer. For instance, consider the function

$$f(x) = -\cos x + \varepsilon x^2 e^{-x^2}, \quad x \in \mathbb{R}.$$

This function has a unique global minimizer $x = 0$, and the global minimum $m = -1$ for any fixed value of $\varepsilon > 0$. If $g_\delta(x)$ is a continuous function, such that

$$\sup_{x \in \mathbb{R}} |g_\delta(x)| \leq \delta, \quad g_\delta(0) > 0,$$

then one can choose $g_\delta(x)$ so that the global minimizer will be as far from $x = 0$ as one wishes.

Example 2.1.8. *Inverse scattering problem in quantum mechanics [R44]*
Let

$$[\nabla^2 + k^2 - q(x)]u = 0 \quad \text{in } \mathbb{R}, \quad k = \text{const} > 0, \tag{2.1.23}$$

$$u = e^{ik\alpha\cdot x} + v, \quad \alpha \in S^2, \tag{2.1.24}$$

$$\lim_{r\to\infty} \int_{|x|=r} |\frac{\partial v}{\partial |x|} - ikv|^2 ds = 0, \tag{2.1.25}$$

where S^2 is the unit sphere in $\mathbb{R}^3$, α is a given unit vector, the direction of the incident plane wave, q is a real-valued function, which is called potential, and which we assume compactly supported,

$$q \in Q_a := \{q : q(x) = 0 \text{ if } |x| > a, \quad q(x) \in L^2(B_a), \quad q = \bar{q}\}, \tag{2.1.26}$$

where $B_a = \{x : |x| \le a\}$. One can prove that the scattered field v is of the form

$$v = A(\alpha', \alpha, k)\frac{e^{ikr}}{r} + o\left(\frac{1}{r}\right), \quad r := |x| \to \infty, \quad \frac{x}{r} = \alpha'. \tag{2.1.27}$$

The coefficient $A = A(\alpha', \alpha, k)$ is called the scattering amplitude. The scattering problem (2.1.23)-(2.1.25) is uniquely solvable under the above assumptions (and even under less restrictive assumptions on the rate of decay of q at infinity, see e.g. [R44]). Therefore, the scattering amplitude $A = A_q$ is uniquely determined by the potential q. The inverse scattering problem of quantum mechanics consists of finding the potential from the knowledge of the scattering amplitude A on some subset of $S^2 \times S^2 \times \mathbb{R}_+$.

A detailed discussion of this problem in the case when the above subset is $S_1^2 \times S_2^2 \times k_0, k_0 = \text{const} > 0$, and S_1^2 and S_2^2 are arbitrary small open subsets of S^2, that is, in the case of fixed-energy data, is given in [R44]. The inverse scattering problem, formulated above, is ill-posed: a small perturbation of the scattering amplitude may be a function $A(\alpha', \alpha, k_0)$ which is not a scattering amplitude corresponding to a potential from the class Q_a, or even from a larger class of potentials.

The author [R16, R17] has established the uniqueness of the solution to inverse scattering problem with fixed energy data, gave a characterization of the class of functions which are the scattering amplitudes at a fixed energy of a potential $q \in Q_a$, and gave an algorithm for recovery of a q from $A(\alpha', \alpha) := A(\alpha', \alpha, k_0)$ known for all $\alpha \in S^2$ and all $\alpha' \in S^2$ at a fixed $k = k_0 > 0$ (see also [R44]).

The error of this algorithm is also given in [R44], see also [R31]. Also a stable estimate of a $q \in Q_a$ is obtained in [R44] when the noisy data $A_\delta(\alpha', \alpha)$ are given,

$$\sup_{\alpha,\alpha' \in S^2} |A_\delta(\alpha', \alpha) - A(\alpha', \alpha)| \leq \delta. \tag{2.1.28}$$

Recently ([R65]) the author has formulated and solved the following inverse scattering-type problem with fixed $k = k_0 > 0$ and fixed $\alpha = \alpha_0$ data $A(\beta) := A(\beta, \alpha_0, k_0)$, known for all $\beta \in S^2$. The problem consists in finding a potential $q \in L^2(D)$, such that the corresponding scattering amplitude $A_q(\beta, \alpha_0, k_0) := A(\beta)$ would approximate an arbitrary given function $f(\beta) \in L^2(S^2)$ with arbitrary accuracy:

$$||f(\beta) - A(\beta)||_{L^2(S^2)} < \epsilon,$$

where $\epsilon > 0$ is an a priori given, arbitrarily small, fixed number. In [R65] it is proved that this problem has a (non-unique) solution, and an analytic formula is found for one of the potentials, which solve this problem. The domain $D \in \mathbb{R}^3$ in the above problem is an arbitrary bounded domain.

Example 2.1.9. *Inverse obstacle scattering.*

Consider the scattering problem:

$$(\nabla^2 + k^2)u = 0 \text{ in } D' := \mathbb{R}^3 \setminus D, \tag{2.1.29}$$

$$u|_S = 0 \tag{2.1.30}$$

$$u = e^{ik\alpha\cdot x} + A(\alpha', \alpha)\frac{e^{ikr}}{r} + o\left(\frac{1}{r}\right), \quad r := |x| \to \infty, \quad \alpha' = \frac{x}{r}, \tag{2.1.31}$$

where D is a bounded domain with boundary S, $k = const > 0$ is fixed, $\alpha \in S^2$ is given, and the coefficient $A(\alpha', \alpha)$ is called the scattering amplitude.

Existence and uniqueness of the solution to problem (2.1.29)-(2.1.31), where D is an arbitrary bounded domain is proved in [R44], where some references concerning the history of this problem can be found. In [R44] one also finds proofs of the existence and uniqueness of the solution to similar problems with boundary conditions of Neumann type

$$u_N|_S = 0, \tag{2.1.32}$$

where N is the unit exterior normal to the surface S,

and of Robin type

$$(u_N + hu)|_S = 0, \tag{2.1.33}$$

under minimal assumptions on the smoothness of the boundary S. In (2.1.33) $h \geq 0$ is an $L^\infty(S)$ function. If the Neumann conditions holds, then S is assumed to be such that the imbedding operator

$$i_1 : H^1(D'_{\mathbb{R}}) \to L^2(D'_{\mathbb{R}})$$

is compact. Here D'_1 is an open subset of D', $D'_1 = D' \cap B_R$, where B_R is some ball containing D.

If the Robin condition holds, then we assume that i_1 and i_2 are compact, where i_1 has been defined above and

$$i_2 : H^1(D'_{\mathbb{R}}) \to L^2(S).$$

Here $L^2(S)$ is the L^2 space with the Hausdorff $(n-1)$-measure on it. The Hausdorff d-measure (d-dimensional measure) is defined as follows. If S is a set in $\mathbb{R}^n$, consider various coverings of this set by countably many balls of radii $r_j \leq r$. Let

$$h(r) := B(d) \inf \sum_j r_j^d,$$

where $B(d)$ is the volume of a unit ball in $\mathbb{R}^d$ and the infimum is taken over all the coverings of S . Clearly $h(r)$ is a non-increasing function, so that it is nondecreasing as $r \to 0$. Therefore there exists the limit (finite or infinite)

$$\lim_{r \to 0} h(r) := \Lambda(S).$$

This limit $\Lambda(S)$ is called d-dimensional Hausdorff measure of S. The restriction on the smoothness of S, which are implied by the compactness of the imbedding operators i_1 and i_2, are rather weak: any Lipschitz boundary S satisfies these restrictions, but Lipschitz boundaries form a small subset of the boundaries for which i_1 and i_2 are compact. (see [GR1, GR]).

The existence and uniqueness of the solution to the obstacle scattering problem imply that the scattering amplitude $A(\alpha', \alpha)$ is uniquely defined by the boundary S and by the boundary condition on S (the Dirichlet condition (2.1.30), the Neumann condition (2.1.32), or the Robin one (2.1.33)).

The inverse obstacle scattering problem consists of finding S and the boundary condition (the Dirichlet, Neumann, or Robin) on S given the scattering amplitude on a subset of $S^2 \times S^2 \times \mathbb{R}_+$. The first basic uniqueness theorem for this inverse problem has been obtained by M. Schiffer in 1964 (see [R13, R44], M.Schiffer did not publish his beautiful proof). He assumed that the Dirichlet condition (2.1.30) holds and that $A(\alpha', \alpha, k)$ is known for a fixed $\alpha = \alpha_0$, all $\alpha' \in S$ and all $k > 0$.

The second basic uniqueness theorem has been obtained in 1985 ([R13]) by the author, who did not preassume the boundary condition on S and proved the following uniqueness theorem:

The scattering data $A(\alpha', \alpha)$, given at an arbitrary fixed $k = k_0 > 0$ for all $\alpha' \in S_1^2$ and $\alpha \in S_2^2$, determine uniquely the surface S and the boundary condition on S of Dirichlet, Neumann, or Robin type.

Here S_1^2 and S_2^2 are arbitrarily small fixed open subsets of S^2 (solid angles), and the boundary condition is either Dirichlet, or Neumann, or Robin type. It is still an open problem to prove the uniqueness theorem for inverse obstacle scattering problem if $A(\alpha') := A(\alpha', \alpha_0, k_0)$ is known for all $\alpha' \in S^2$, a fixed $\alpha = \alpha_0 \in S^2$ and a fixed $k = k_0 > 0$.

A recent result ([R51] in this direction is a uniqueness theorem under additional assumptions on the geometry of S (convexity of S and nonanalyticity of S).

The inverse obstacle scattering problem is ill-posed by the same reason as the inverse potential scattering problem in example 2.1.8: small perturbation of the scattering amplitude may throw it out of the set of scattering amplitudes. A characterization of the class of scattering amplitudes is given in ([R15], see also [R14, R19]).

The absolute majority of the practically interesting inverse problems are ill-posed.

Let us mention some of these problems in addition to the two inverse scattering problems mentioned above.

Example 2.1.10. *Inverse problem of geophysics.*

Let

$$[\nabla^2 + k^2 + k^2 v(x)]u = -\delta(x - y) \text{ in } \mathbb{R}^3, \tag{2.1.34}$$

where $k = \text{const} > 0$, $v(x)$ is a compactly supported function, $v \in L^2(D)$, $\bar{D} = \text{supp } v \subset \mathbb{R}^3_-$, where supp v is the support of v, and $\mathbb{R}^3_- := \{x : x_3 < 0\}$. We assume that u satisfies the radiation condition

$$\frac{\partial u}{\partial |x|} - iku = o\left(\frac{1}{|x|}\right), \quad |x| \to \infty, \tag{2.1.35}$$

uniformly in directions $\frac{x}{|x|}$. One may think that $P = \{x : x_3 = 0\}$ is the surface of the earth, $v(x)$ is the inhomogeneity in the velocity profile, u is the acoustic pressure, and y is the position of the point source of this pressure.

The simplest model inverse problem of geophysics consists of finding $v(x)$ from the knowledge of $u(x, y, k)$ for all $x \in P_1$, all $y \in P_2$ and a fixed $k > 0$, (or for all $k \in (0, k_0)$, where $k_0 > 0$ is arbitrarily small fixed number; in this case the data $u(x, y, k)$, $k \in (0, k_0)$, are called low-frequency data).

Here P_1 and P_2 are open sets in P . A more realistic model allows one to replace equation (2.1.31) with

$$[\nabla^2 + k^2 n(x) + k^2 v(x)]u = \delta(x-y) \text{ in } \mathbb{R}, \tag{2.1.36}$$

where the non-constant background refraction coefficient $n(x)$ is known. It can be fairly arbitrary function ([R19]).

In geophysical modeling one often assumes that $n_0(x) = 1$ for $x_3 > 0$ (in hot air) and $n(x) = n_0 = \text{const}$ for $x_3 < 0$ (homogeneous earth).

The inverse geophysical problem is ill-posed: a small perturbation of the data $u(x,y,k)$, $x, y \in P_1 \times P_2$ may lead to a function which is not the value of the solution to problem (2.1.34)-(2.1.35) for a $v \in L^2(D)$.

Example 2.1.11. *Finding small subsurface inhomogeneities from surface scattering data.*

The inverse problem can be formulated as the inverse problem of geophysics in Example 2.1.10, with the additional assumptions

$$D = \cup_{j=1}^J D_j, \quad \text{diam} D_j := a_j, \quad \max_j a_j := a_j, \quad ka \ll 1, \quad kd \gg 1, \tag{2.1.37}$$

where

$$d = \min_{i \neq j} \text{dist}(D_i, D_j).$$

The inverse problem is:

Given $u(x,y,k)$ for $x \in P_1, y \in P_2$, where P_1 and P_2 are the same as in Example 2.1.10, and $k = k_0 > 0$ fixed, find the positions of D_j, their number J, and their intensities $V_j := \int_{D_j} v(x)dx$.

This problem is ill-posed by the same reasons as the inverse geophysical problem. a method for solving this problem is given in [R47].

Example 2.1.12. *Antenna synthesis problem.*

Let an electric current be flowing in a region D. This current creates an electromagnetic field according to the Maxwell's equations

$$\nabla \times E = i\omega\mu H, \quad \nabla \times H = -i\omega\varepsilon E + j, \tag{2.1.38}$$

where ω is the frequency, ε and μ are dielectric and magnetic parameters, and j is the current. If ε and μ are constants, then one can derive the following formula ([R44], p.11).

$$E = -i\omega\mu \frac{e^{ikr}}{r}[\alpha', [\alpha', J]] + o\left(\frac{1}{r}\right), \quad r = |x|, \quad \alpha' = \frac{x}{r}, \tag{2.1.39}$$

where $k = \omega\sqrt{\varepsilon\mu}$, $[a, b]$ is the cross product, and

$$J = \frac{1}{4\pi}\int_D e^{-ik\alpha'\cdot y} j(y)dy, \tag{2.1.40}$$

where the integral is taken over the support D of the current $j(y)$. We assume that D is bounded.

The inverse problem, which is called the antenna synthesis problem, consists of finding $j(x)$ from the knowledge of the radiation pattern

$$[\alpha', [\alpha', J]]$$

for all $\alpha' \in S^2$.

This problem is ill-posed and, in general, it may have many solutions. One has to restrict the admissible currents to obtain an inverse problem which has at most one solution. For example, let us assume that

$$j(x) = j(x_3)\delta(x_1)\delta(x_2)e_3, \quad x_3 := z, \quad -a \leq z \leq a,$$

where $\delta(x_j)$ is the delta-function and e_3 is the unit vector along the z-axis. Thus, we deal with the linear antenna. The radiation pattern in this case is

$$\frac{1}{4\pi}[\alpha', [\alpha', e_3]] \int_{-a}^{a} e^{-ikzu} j(z)dz, \quad u = cos\theta = e_3 \cdot \alpha',$$

and the problem of linear antenna synthesis consists of finding $j(z)$ from the knowledge of the desired diagram $f(u)$:

$$\int_{-a}^{a} e^{-ikzu} j(z)dz = f(u), \quad -1 \leq u \leq 1. \tag{2.1.41}$$

Equation (2.1.41) has at most one solution. Indeed, if $f(u) = 0$, then the left-hand side of (2.1.41) vanishes on the set $-1 \leq u \leq 1$ and is an entire function of u.

By the uniqueness theorem for analytic functions, one concludes that

$$\int_{-a}^{a} e^{-ikzu} j(z)dz = 0 \quad \text{for all } u \in \mathbb{C},$$

and, in particular, for all $u \in \mathbb{R}$. This and the injectivity of the Fourier transform imply $j(z) = 0$. So, the uniqueness of the solution to (2.1.41) is proved. Solving equation (2.1.41) is an ill-posed problem because small perturbations of f may be non-analytic, and equation (2.1.41) with any analytic right-hand side f has no solution. There is an extensive literature on antenna synthesis problems ([AV, MY, R2, R3, R5, R6]).

Example 2.1.13. *Inverse problem of potential theory.*

Consider the gravitational potential generated by a mass with density ρ distributed in a domain D:

$$\int_D \frac{\rho(y)dy}{4\pi|x-y|} = u(x). \tag{2.1.42}$$

The inverse problem of potential theory consists of finding $\rho(y)$ from the measurements of the potential $u(x)$ outside D. This problem is not uniquely solvable, in general. For example, a point mass distribution, $\rho(y) = M\delta(x)$ generates the potential $\frac{M}{4\pi|x|}$ in the region $|x| > 0$, which is the same as the mass the uniformly distributed in a ball $B_a = \{x : |x| \leq a\}$ with the density

$$\rho(y) = \frac{3M}{4\pi a^3},$$

so that the total mass of the ball is equal to M.

However, if one assumes a priori that the mass density $\rho = 1$, then the domain D, which is star-shaped with respect to a point $x \in D$, is uniquely determined by the knowledge of the potential $u(x)$ in the region $|x| > R$, where the ball B_R contains D

Example 2.1.14. *Tomography and integral geometry problems.*

Suppose there is a family of curves, and the integrals of a function over every curve from this family are known. The problem consists of recovery of f from the knowledge of these integrals. An important example is tomography. If the family of straight lines $l_{\alpha p}$, where

$$l_{\alpha p} = \{x : \alpha \cdot x = p\}, \quad \alpha \subset S^1,$$

where S^1 is the unit sphere in $\mathbb{R}^2$, $x \in \mathbb{R}^2$, $p \geq 0$ is a number, then the knowledge of the family of integrals

$$\hat{f}(\alpha, p) = \int_{l_{\alpha p}} f(x)ds, \tag{2.1.43}$$

allows one to recover f uniquely. Analytical inversion formulas are known ([R25]). The function $\hat{f}(\alpha, p)$ is called the Radon transform of f. In applications $\hat{f}(\alpha, p)$ is called a tomogram. Finding f from $\hat{f}$ is an ill-posed problem: if $\hat{f}$ is the Radon transform of a compactly supported continuous function f (or L^2 function f), and if $\hat{f}$ is perturbed a little in the sup-norm (or L^2-norm) then the resulting function $\hat{f} + h$, $||h|| \leq \delta$, may be not the Radon transform of any compactly supported L^2-function.

In practice a dominant role is played by the local tomography.

In local tomography one has only the local tomographic data, i.e., the values $\hat{f}(\alpha, p)$ for α, p which satisfy the inequality:

$$|\alpha \cdot x - p| \leq \varepsilon, \tag{2.1.44}$$

where $x_0 \in \mathbb{R}^2$ is a given fixed point in a region of interest, for example, in a region of human body around which one thinks a tumor is possible, and $\varepsilon > 0$ is a small number. One of the basic motivations for the theory of local tomography, originated in [R24, R23], and developed in [R25], was the desire to minimize the X-ray radiation of patients. From the knowledge of local tomographic data one cannot find the function f, because the inversion formula in $\mathbb{R}^2$ is nonlocal (e.g., see [R25], p.31):

$$f(x) = \frac{1}{4\pi} \int_{S^1} d\alpha \int_{-\infty}^{\infty} \frac{\hat{f}_p(\alpha, p)dp}{\alpha \cdot x - p}. \tag{2.1.45}$$

Here

$$\hat{f}(\alpha, p) = \hat{f}(-\alpha, -p) \quad \forall p \in \mathbb{R},$$

S^1 is the unit sphere in $\mathbb{R}^2$, i.e. the set $|\alpha| = 1, \quad \alpha \in \mathbb{R}^2$, and $\hat{f}_p = \frac{\partial \hat{f}}{\partial p}$. Formula (2.1.45) is proved in [R25] for smooth rapidly decaying functions f, but it remains valid for $f \in H_0^1(\mathbb{R}^2)$, where $H_0^1(\mathbb{R}^2)$ is the set of functions in the Sobolev space $H^1(\mathbb{R})$ with compact support.

Nonlocal nature of the inversion formula (2.1.45) means that one has to know $\hat{f}(\alpha, p)$ for all $\alpha \in S^1$ and all $p \in \mathbb{R}$ in order to recover $f(x)$ at the point x. The author has posed the following question:

If one cannot recover $f(x)$ from local tomographic data, what practically useful information can one recover from these data?

The answer, given in ([R23, R24, R25]), is:

One can recover the discontinuity curves of f and the sizes of the jumps of f across the discontinuity curves.

Example 2.1.15. *Inverse spectral problem.*

Consider the problem

$$lu := -u'' + q(x)u + \lambda u, \quad 0 \leq x \leq 1, \tag{2.1.46}$$

$$u(0) = u(1) = 0. \tag{2.1.47}$$

Assume that

$$q = \bar{q}, \quad q \in L^1(0,1). \tag{2.1.48}$$

Then problem (2.1.46)-(2.1.47) has a discrete spectrum

$$\lambda_1 < \lambda_2 \leq ..., \quad \lim_{n\to} \lambda_n = \infty. \tag{2.1.49}$$

A natural question is:

Does the set of eigenvalues $\{\lambda_j\}_{j=1,2,...}$ determine $q(x)$ uniquely?

The answer is:

It does not, in general.

The set of $\{\lambda_j\}_{\forall j}$ determines roughly speaking half of the potential $q(x)$ in the following sense (see, e.g., [R44]): if $q(x)$ is unknown on the half of the interval $0 \leq x \leq \frac{1}{2}$, then the knowledge of all $\{\lambda_j\}_{\forall j}$ determines $q(x)$ on the remaining half of the interval $\frac{1}{2} < x < 1$ uniquely.

There is an exceptional result, however, due to Ambarzumian (1929), which says that if $u'(0) = u'(1) = 0$, then the set of the corresponding eigenvalues $\{\mu_j\}_{\forall j}$ determines q uniquely ([R44]). A multidimensional generalization of this old result is given in [R44].

Let us define the spectral function of the selfadjoint Dirichlet operator $l = -\frac{d^2}{dx^2} + q(x)$ in $L(\mathbb{R}_+)$, $\mathbb{R}_+ = [0, \infty)$. This operator can be defined as the closure of the symmetric operator l_0 defined on twice continuously differentiable functions u vanishing at $x = 0$ and near infinity and such that $lu \in L(\mathbb{R}_+)$.

It is not trivial to prove that l_0 is densely defined in $H = L^2(\mathbb{R}_+)$ if one assumes that $q \in L^1(\mathbb{R}_+)$ or if

$$q \in L_{1,1} := \{q : q = \bar{q}, \quad \int_0^\infty x|q(x)|dx < \infty\}.$$

The idea of the proof is as follows (cf. [N]): the operator

$$l_0 + a^2 = \frac{d^2}{dx^2} + q(x) + a^2$$

is symmetric on its domain of definition $D(l_0) \subset H$, as one can check easily by integration by parts; the equation

$$lu + a^2u = -u'' + q(x)u + a^2u = f, \quad u(0) = 0, \tag{2.1.50}$$

is uniquely solvable for any sufficiently large $a > 0$ and for any $f \in H$, and its solution $u \in H^1(\mathbb{R}_+)$; if $h \in H$ and $h \perp D(l_0)$, then $h = lv$ and $(v, l_0w + a^2w) = 0 \;\; \forall w \in D(l_0)$, where (u, w) is the inner product in $L^2(0, \infty)$; consequently $v \in H$ solves homogeneous ($f = 0$) problem (2.1.50); if $a > 0$ is sufficiently large this problem has only the initial the trivial solution $v = 0$; thus $h = 0$, and the claim is proved: $D(l_0)$ is dense in H.

With the selfadjoint Dirichlet operator l one associates the spectral function $d\rho(\lambda)$. This is the function for which

$$\int_0^\infty |f(x)|^2 dx + \sum_{j=1}^{J} (f, \varphi_j)\varphi_j = \int_{-\infty}^{\infty} |\tilde{f}(\lambda)|^2 d\rho(\lambda), \tag{2.1.51}$$

for any $f \in L^2(\mathbb{R}_+)$.

Here

$$\tilde{f}(\lambda) = \int_0^\infty f(x)\varphi(x,\lambda)dx, \quad l\varphi = \lambda\varphi, \quad \varphi(0,\lambda) = 0, \quad \varphi'(0,\lambda) = 1,$$

and

$$l\varphi_j = \lambda_j\varphi_j, \quad \varphi_j(0) = 0, \quad ||\varphi_j||_H = 1, \tag{2.1.52}$$

i.e. φ_j are the normalized eigenfunctions of l, corresponding to possibly existing negative spectrum of l. If $q \in L_{1,1}$ this spectrum is known to be finite (e.g. see [G]).

If $q \in L_{1,1}$ then there is a unique spectral function corresponding to the selfadjoint operator l.

The inverse spectral problem consists of finding $q(x)$ from the knowledge of $\rho(\lambda)$. This problem is ill-posed: small perturbations of $\rho(\lambda)$ may lead to a function which is not a spectral function.

Example 2.1.16. *Inverse problems for wave equation.*

Consider the wave equation

$$\frac{1}{c^2(x)} u_{tt} = \Delta u + \delta(x - y) \text{ in } \mathbb{R}^3, \tag{2.1.53}$$

$$u|_{t=0} = 0, \quad u_t|_{t=0} = 0. \tag{2.1.54}$$

The function $c(x)$ is the wave velocity. Assume that

$$c^{-2}(x) = c_0^{-2}(x)[1 + v(x)],$$

where $c_0(x)$ is the wave velocity in the background medium and $v(x)$ is the inhomogeneity in the wave velocity. Assume that $v(x)$ is unknown, compactly supported in $\mathbb{R}^3_- := \{x : x_3 < 0\}$.

The inverse problem is:

Given the measurements of $u(x,y,t)$, the solution to (2.1.53)-(2.1.54) for all values of $x \in P_1$, $y \in P_2$ and $t \in [0,T]$, where $T > 0$ is some number, find $v(x)$, provided that $c_0(x)$ is known. Here P_1 and P_2 are open sets on the plane $x_3 = 0$, which is the surface of the Earth in the geophysical model.

This inverse problem is ill-posed: small perturbations of $u(x,y,t)$ may lead to a function which is not a restriction of the solution to problem (2.1.53)-(2.1.54) to the plane $x_3 = 0$.

Example 2.1.17. *Inverse problems for the heat equation.*

Consider the problem

$$u_t = \Delta u - q(x)u, \quad t \geq 0, \quad x \in D, \tag{2.1.55}$$

$$u|_{t=0} = 0, \tag{2.1.56}$$

$$u|_S = f, \quad S := \partial D. \tag{2.1.57}$$

This problem has a unique solution $u(x,t)$.

Let the flux

$$u_N|_S = h(s,t), \tag{2.1.58}$$

be measured for any $f \in H^{\frac{3}{2}}(S)$, where $H^m(S)$ is the Sobolev space and N is the exterior unit normal to S.

The inverse problem is:

Given the set of pairs $\{f, h\}$ for all $t \in [0,T]$, where $T > 0$ is a number, find find $q(x)$.

This problem is ill-posed: small perturbations of h may lead to a function which is not a normal derivative of a solution to problem of the type (2.1.55)-(2.1.57).

Example 2.1.18. *Inverse conductivity problem.*

This problem is also called impedance tomography problem. Consider the stationary problem

$$\nabla \cdot (a(x)\nabla u) = 0 \text{ in } D, \quad u|_S = f, \tag{2.1.59}$$

where

$$0 < a_0 \leq a(x) \leq a_1 < \infty, \quad a \in H^2(D), \tag{2.1.60}$$

and assume that $S = \partial D$ is sufficiently smooth. Problem (2.1.59) has a unique solution, so the function

$$u_N|_S = h, \tag{2.1.61}$$

where N is the unit exterior normal to S, is uniquely defined if $a(x)$ and f are known.

The inverse problem is:

Given the set $\{f, h\}$, find $a(x)$.

This problem has at most one solution but it is ill-posed (see e.g. [R19] and [R44]).

Example 2.1.19. *Deconvolution and other imaging problems.*
Consider, for instance, the problem

$$\int_0^t k(t,s)u(s)ds = f(t). \tag{2.1.62}$$

The deconvolution problem consists of finding $u(s)$ from the knowledge of f and $k(t,s)$. This problem is ill-posed as was explained in Example 2.1.4, where a more general Fredholm-type equation of the first kind was discussed.

Example 2.1.20. *Heat equation with reversed time.*
Consider the problem:

$$u_t = u_{xx}, \quad t \geq 0, \quad 0 \leq x \leq x_b; \quad u(0,t) = u(x_b,t) = 0, \quad u(x,0) = f(x).$$

This problem has a unique solution:

$$u(x,t) = \sum_{j=1}^{\infty} e^{-\lambda_j t} f_j \varphi_j(x); \quad f = (f, \varphi_j)_{L^2(0,\pi)} := (f, \varphi_j),$$

$$\varphi_j(x) = \sqrt{\frac{2}{\pi}} sin(jx); \quad \lambda_j = j^2; \quad (\varphi_j, \varphi_m) = \delta_{j,m}.$$

Consider the following problem:
Given the function $g(x)$ and a number $T > 0$, can one find $f(x) = u(x,0)$ such that $u(x,T) = g(x)$?

In general, the answer is no: not every $g(x)$ can be the value of the solution $u(x,t)$ of the heat equation at $t = T > 0$. for example, $g(x)$ has to be infinitely differentiable. However, given an arbitrary $g \in L^2(0,\pi)$ and an arbitrary small $\varepsilon > 0$, one can find f, such that $||u(x,T) - g(x)|| \leq \varepsilon$, where $u(x,t)$ is the solution to the heat equation with $u(x,0) = f(x)$. This is easy to see from the formula

$$u(x,T) = \sum_{j=1}^{\infty} e^{-\lambda_j T} f_j \varphi_j(x).$$

The inequality

$$||u(x,T) - g|| \leq \varepsilon$$

holds if

$$\sum_{j=1}^{\infty} |e^{-\lambda_j T} f_j - g_j|^2 \leq \varepsilon^2.$$

If $\sum_{j\geq j(\varepsilon)} |g_j|^2 < \frac{\varepsilon^2}{2}$, then one may take

$$f_j = 0 \quad \text{for} \quad j \geq j(\varepsilon),$$

and

$$f_j = g_j e^{\lambda_j T} \quad \text{for} \quad j < j(\varepsilon),$$

and get

$$||u(x,T) - g|| \leq \varepsilon.$$

Note that if $T > 0$ and $j(\varepsilon)$ are large, then the coefficients f_j are extremely large.

Thus, the above problem is highly ill-posed: small perturbations of g may lead to arbitrary large perturbations of f, or throw g out of the set of functions, which are the values of $u(x,T)$. Similar results hold in multidimensional problems for the heat equation.

The number of examples of ill-posed problems, which are of interest in applications, can be easily increased.

The number of examples of ill-posed problems, which are of interest in applications, can be easily increased.

Let us define the notion of regularizer.

Definition 2.1.2. *An operator R_δ is called a regularizer for the problem*

$$Au = f$$

if

$$\lim_{\delta\to 0} ||R_\delta f_\delta - u|| = 0, \tag{2.1.63}$$

where R_δ is a bounded operator defined on the whole space, the relation (2.1.63) should hold for any $f \in R(A)$ and some solution u of equation (2.1.63).

In the literature it is often assumed that A is injective, and then the solution to equation (2.1.63) is unique. However, one may also consider the case when A is not injective. Usually the operator R_δ is constructed as a two-parameter family $R_{\delta,a}$, where parameter a is called a regularization parameter , and for a suitable choice of $a = a(\delta)$ the operator $R_{\delta,a(\delta)} := R_\delta$ satisfies Definition 2.1.2 methods for constructing of $R_{\delta,a}$ and for choosing $a = a(\delta)$ are discussed in many papers and books (e.g. see [I], [R44] [VA], and references therein).

Sometimes the requirement (2.1.63) is replaced by

$$\lim_{\delta\to 0} \sup_{f_\delta \in B(f,\delta)} ||R_\delta f_\delta - u|| = 0, \tag{2.1.64}$$

where $B(f,\delta) = \{g : ||g - f|| \le \delta\}$.

The author has proposed a different definition of the regularizer R_δ (see [R44]):

Definition 2.1.3. *An operator R_δ is a regularizer for the problem $Au = f$ if*

$$\lim_{\delta\to 0} \sup_{f \in B(f_\delta,\delta),\ f=Au} ||R_\delta f_\delta - u|| = 0, \tag{2.1.65}$$

where $B(f_\delta,\delta) = \{g : ||g - f_\delta|| \le \delta\}$.

The motivation for this new definition is natural: the given data are $\{f_\delta, \delta\}$ and f is unknown. This unknown f may be any element of the set $B(f_\delta,\delta) \cap R(A)$. Therefore the regularizer must recover u corresponding to any such f. There is a considerable practical difference between the two definitions (2.1.64) and (2.1.65). For instance, one may be able to find a regularizer in the sense (2.1.64) but this regularizer will not be a regularizer in the sense (2.1.65). Consider a particular example. Let

$$R_\delta f_\delta = \frac{f_\delta(x + h(\delta)) - f_\delta(x - h(\delta))}{2h(\delta)}. \tag{2.1.66}$$

This is a regularizer for the problem of stable numerical differentiation. It was first proposed in [R4] and then used extensively (see [R44]). The choice of $h(\delta)$ depends on the a priori information about the unknown function f whose derivative we want to estimate stably given the data $\{f_\delta, \delta, M_a\}$, where $||f_\delta - f||_\infty \le \delta$, the norm is $L^\infty(a,b)$-norm, the interval (a,b) is arbitrary. Without loss of generality one may take $a = 0$, $b = 1$, which we will do below. The number M_a is the a priori known upper bound for the derivative of f of order $a > 0$. If $a = j + b$, where j is an integer, $j > 0$, and $b \in (0,1)$, then

$$M_a = \sup_{x\in[0,1]} \{|f^{(j)}(x)| + |f(x)|\} + \sup_{x,y\in[0,1]} \frac{|f^{(j)}(x) - f^{(j)}(y)|}{|x-y|^b}. \tag{2.1.67}$$

It is proved in [R44] that no regularizer (linear or nonlinear) can be found for the problem of stable numerical differentiation of noisy data if the regularizer is understood in the sense (2.1.65) and $a \le 1$, and that such a regularizer can be found in the form (2.1.66) with a suitable $h(\delta)$ provided that $a > 1$. A regularizer in the sense (2.1.64) can be found for the problem of stable numerical differentiation with $a = 1$, however practically this regularizer is of no use. A detailed discussion of this problem is given in Section 15.2.

2.2 Variational regularization

Consider a linear ill-posed problem

$$Au = f, \tag{2.2.1}$$

where A is a closed, densely defined linear operator in a Hilbert space H.

We assume that problem (2.2.1) is ill-posed, that $Ay = f$, where

$$y \perp \mathcal{N}, \quad \mathcal{N} = \mathcal{N}(A) = \{u : Au = 0\},$$

and that f_δ is given in place of f,

$$||f_\delta - f|| \leq \delta, \quad ||f_\delta|| > c\delta, \quad c = \text{const}, \quad c \in (1, 2).$$

$$\mathcal{F}(u) := ||Au - f_\delta||^2 + a||u||^2 = \min, \tag{2.2.2}$$

where $a > 0$ is regularization parameter.

The method of variational regularization for stable solution of equation (2.2.1), that is, for finding the regularizer R_δ such that (2.1.63) holds, consists of finding the functional (2.2.2) and then choosing $a = a(\delta)$ such that

$$\lim_{\delta \to 0} a(\delta) = 0, \tag{2.2.3}$$

and (2.1.63) holds with $R_\delta f_\delta = u_\delta := u_{a(\delta),\delta}$.

Let us show that the global minimizer of (2.2.2) exists and is unique. Indeed, functional (2.2.2) is quadratic, so a necessary condition for its minimizer is a linear equation, the Euler equation. Assume first that A is bounded. Then the Euler equation for the functional (2.2.2) is

$$A^*Au + au = A^*f_\delta. \tag{2.2.4}$$

Let us denote

$$A^*A := T, \quad T_a : T + aI. \tag{2.2.5}$$

Since $T = T^* \geq 0$, the operator T has a bounded inverse, $||T_a^{-1}|| \leq \frac{1}{a}$, so equation (2.2.4) has a solution $u_{a,\delta} = T_a^{-1}A^*f_\delta$ and this solution is unique. One has

$$\mathcal{F}(u_{a,\delta}) \leq \mathcal{F}(u). \tag{2.2.6}$$

Indeed, one can check that

$$\mathcal{F}(u_{a,\delta}) \leq \mathcal{F}(u_{a,\delta} + v) \quad \forall v \in H,$$

and the equation sign is attained if and only if $v = 0$.

Let us choose $a = a(\delta)$ so that $u_\delta := u_{a(\delta),\delta}$ would satisfy the relation:

$$\lim_{\delta \to 0} ||u_\delta - y|| = 0. \tag{2.2.7}$$

To do so, we estimate:

$$||T_a^{-1}A^*f_\delta - y|| \leq ||T_a^{-1}A^*(f_\delta - f)|| + ||T_a^{-1}A^*f - y|| := J_1 + J_2. \tag{2.2.8}$$

Note that

$$||T_a^{-1}A^*|| \leq \frac{1}{2\sqrt{a}}. \tag{2.2.9}$$

To prove (2.2.9) one uses the commutation formula

$$T_a^{-1}A^* = A^*Q_a^{-1}, Q := AA^*. \tag{2.2.10}$$

This formula is easy to check: multiply (2.2.10) by Q_a from the right and by T_a from the left and get an obvious relation:

$$A^*(AA^* + aI) = (A^*A + aI)A^*.$$

Reversing the above derivation, one gets (2.2.10). Using the polar decomposition

$$A^* = U(AA^*)^{\frac{1}{2}}, \tag{2.2.11}$$

where U is a partial isometry, one gets, using the spectral theorem,

$$||T_a^{-1}A^*|| = ||A^*Q_a^{-1}|| \leq ||Q^{\frac{1}{2}}Q_a^{-1}|| = \sup_{s \geq 0} \frac{s^{\frac{1}{2}}}{s + a} = \frac{1}{2\sqrt{a}}, \tag{2.2.12}$$

so formula (2.2.9) is verified. Thus

$$J_1 \leq \frac{\delta}{2\sqrt{a}}. \tag{2.2.13}$$

To estimate J_2 in (2.2.8) one uses the spectral theorem again and gets:

$$J_2^2 = ||T_a^{-1}Ty - y||^2 = a^2||T_a^{-1}y|| = \int_0^{||T||} \frac{a^2 d(E_s y, y)}{(a+s)^2} := \beta^2(a). \quad (2.2.14)$$

One has

$$\lim_{a\to 0} \beta^2(a) = ||P_{\mathcal{N}} y||^2 = 0, \quad (2.2.15)$$

because $y \perp \mathcal{N}$ by the assumption, and

$$\mathcal{N} = (E_0 - E_{-0})H.$$

From (2.2.8), (2.2.13)-(2.2.15) one gets:

$$||u_{a,\delta} - y|| \leq \frac{\delta}{2\sqrt{a}} + \beta(a). \quad (2.2.16)$$

Taking any $a(\delta)$ such that

$$\lim_{\delta\to 0} a(\delta) = 0, \quad \lim_{\delta\to 0} \frac{\delta}{\sqrt{a(\delta)}} = 0, \quad (2.2.17)$$

and setting $u_{a(\delta),\delta} := u_\delta$, one obtains (2.2.7).

Let us summarize our result.

Theorem 2.2.1. *Assume that $a = a(\delta)$ and (2.2.17) holds. Then the element $u_\delta = T_{a(\delta)}^{-1} A^* f_\delta$ satisfies (2.2.7).*

Remark 2.2.1. Without additional assumptions on y it is impossible to estimate the rate of decay of $\beta(a)$ as $a \to 0$. Therefore it is impossible to get a rate of convergence in (2.2.7): the convergence can be as slow as one wishes for some y. The usual assumption which would guarantee some rate of decay of β and, therefore, of $||u_\delta - y||$ is the following one:

$$y = T^b z, \quad 0 < b \leq 1. \quad (2.2.18)$$

If (2.2.18) holds, then

$$\beta^2(a) = \int_0^{||T||} \frac{a^2 s^{2b} d(E_s z, z)}{(a+s)^2} \leq a^{2b}(1-b)^{2-2b} b^{2b} ||z||^2. \quad (2.2.19)$$

If $||T|| < \infty$, then one can give a rate for $b > 1$ as well, but the rate will be $O(a)$ as $a \to 0$, so that for $b \geq 1$ there is a saturation. The case $||T|| = \infty$ is discussed below separately.

Thus,

$$\beta(a) \leq c||z||a^b, \quad 0 < b \leq 1, \tag{2.2.20}$$

where $c = (1-b)^{1-b}b^b$.

If one minimizes for a small fixed $\delta > 0$ the function

$$\frac{\delta}{2\sqrt{a}} + c||z||a^b = \min \tag{2.2.21}$$

in the region $a > 0$, then one gets

$$a = a(\delta) = c_1 \delta^{\frac{2}{2b+1}}, \quad c_1 = \left(\frac{1}{4bc||z||} \right)^{\frac{2}{2b+1}}. \tag{2.2.22}$$

Thus, under the condition (2.2.20), one gets

$$||u_\delta - y|| = O(\delta^{\frac{2}{2b+1}}), \quad \delta \to 0. \tag{2.2.23}$$

Let us now generalize the above theory to include unbounded, closed, densely defined operators A. The main difficulty is the following one: the element f_δ in (2.2.4) may not belong to the domain $D(A^*)$ of A^*. Our result is stated in the following theorem.

Theorem 2.2.2. *The operator $T_a^{-1}A^*$, defined originally on $D(A^*)$, is closable. Its closure denoted again $T_a^{-1}A^*$ is defined on all of H and is a bounded linear operator with the norm bounded as in (2.2.9) for any $a > 0$. The relation (2.2.10) holds.*

Proof of Theorem 2.2.2. Let us first check that the operator $T_a^{-1}A^*$ with the domain $D(A^*)$ is closable. Recall that a densely defined linear operator B in a Hilbert space is closable if it has a closed extension. This happens if and only if the closure of the graph of B is again a graph. In other words, if $u_n \to 0$ and $Bu_n \to f$, then $f = 0$.

The operator B is called closed if $u_n \to u$ and $Bu_n \to f$ implies $u_n \in D(B)$ and $Bu = f$. The operator B is closed if and only if its graph is a closed subset in $H \times H$, that is, the set $\{u, Bu\}$ is a closed linear subspace in $H \times H$.

Let us check that the operator $T_a^{-1}A^*$ with domain $D(A^*)$ is closable. Let $u_n \in D(A^*)$ and $u_n \to 0$, and assume that $T_a^{-1}A^*u_n \to g$, as $n \to \infty$. We wish to prove that $g = 0$. For any $u \in H$ one gets:

$$(g, u) = \lim_{n\to\infty} (T_a^{-1}A^*u_n, u) = \lim_{n\to\infty} (u_n, AT_a^{-1}u) = 0, \tag{2.2.24}$$

where we have used the closedness of A.

Since u is arbitrary, equation (2.2.24) implies $g = 0$. So the operator $T_a^{-1}A^*$ with the domain $D(A^*)$ is closable.

Estimate (2.2.9) and formula (2.2.10) remain valid and their proofs are essentially the same.

Theorem 2.2.2 is proved. □

Theorem 2.2.3. *The conclusion of Theorem 2.2.1 remains valid for unbounded, closed, densely defined operator A under the same assumptions as in Theorem 2.2.1.*

Proof of Theorem 2.2.3 is the same as that of Theorem 2.2.1 after Theorem 2.2.2 is established.

Let us now discuss an a posteriori choice of $a(\delta)$, the discrepancy principle . Let $u_{a,\delta} = T_a^{-1}A^*f_\delta$. For a fixed $\delta > 0$, consider the equation

$$||Au_{a,\delta} - f_\delta|| = ||AT_a^{-1}A^*f_\delta - f_\delta|| = c\delta, \quad c \in (1,2), \tag{2.2.25}$$

as an equation for $a = a(\delta)$. Here c is a constant and we assume that $||f_\delta|| > c\delta$. We prove the following result.

Theorem 2.2.4. *Equation (2.2.25) has a unique solution $a = a(\delta)$ for every sufficiently small $\delta > 0$, provided that $||f_\delta - f|| \leq \delta$, $\quad ||f_\delta|| > c\delta$, and $f = Ay$. One has $\lim_{\delta\to 0} a(\delta) = 0$ and (2.2.7) holds with $u_\delta := T_{a(\delta)}^{-1}A^*f_\delta$.*

Proof of Theorem 2.2.4. Using formula (2.2.10) and the spectral theorem for the selfadjoint operator Q, one gets

$$||AT_a^{-1}A^*f_\delta - f_\delta||^2 = ||[QQ_a^{-1} - I]f_\delta||^2 = \int_0^\infty \frac{a^2 d(E_s f_\delta, f_\delta)}{(a+s)^2} := h(\delta, a). \tag{2.2.26}$$

The function $h(\delta, a)$ is continuos with respect to a on the interval $(0, \infty)$ for any fixed $\delta > 0$. One has

$$h(\delta, \infty) = \int_0^\infty d(E_s f_\delta, f_\delta) = ||f_\delta||^2 > c^2\delta^2, \tag{2.2.27}$$

and

$$h(\delta, +0) = ||P_{\mathcal{N}^*} f_\delta||^2 \leq \delta^2, \tag{2.2.28}$$

where

$$\mathcal{N}^* = \mathcal{N}(A^*) = \mathcal{N}(Q),$$

P is the orthogonal projector onto $\mathcal{N} = \mathcal{N}(A)$, and we have used the following formulas:

$$\lim_{a\to 0}\int_0^\infty \frac{a^2 d(\mathcal{E}_s f_\delta, f_\delta)}{(a+s)^2} = ||(\mathcal{E}_0 - \mathcal{E}_{-0})f_\delta||^2 = ||P_{\mathcal{N}^*} f_\delta||^2, \tag{2.2.29}$$

where $\mathcal{E}_s$ is the resolution of the identity corresponding to Q, and

$$||P_{\mathcal{N}^*} f_\delta|| \leq ||P_{\mathcal{N}^*} f|| + ||P_{\mathcal{N}^*}(f_\delta - f)|| \leq \delta, \tag{2.2.30}$$

because $P_{\mathcal{N}^*}\mathcal{R}(A) = 0$ and $f \in \mathcal{R}(A)$.

From (2.2.27), (2.2.28) and the continuity of $h(\delta, a)$ it follows that equation (2.2.25) has a solution. This solution is unique because $h(\delta, a)$ is a monotonically growing function of a for a fixed $\delta > 0$.

Also,

$$\lim_{\delta\to 0} a(\delta) = 0,$$

because $\lim_{\delta\to 0} h(\delta, a(\delta)) = 0$, and $h(\delta, a(\delta)) \geq c_1 > 0$ if $\lim_{\delta\to 0} a(\delta) \geq c_2 > 0$, where c_1 and c_2 are some constants.

Let us now check that (2.2.7) holds with $u_\delta = T_a^{-1} A^* f_\delta$. We have

$$\mathcal{F}(u_\delta) = c^2\delta^2 + a(\delta)||u_\delta||^2 \leq \mathcal{F}(y) = \delta^2 + a(\delta)||y||^2. \tag{2.2.31}$$

Since $c > 1$, one gets

$$||u_\delta|| \leq ||y||. \tag{2.2.32}$$

Thus

$$\limsup_{\delta\to 0} ||u_\delta|| \leq ||y||. \tag{2.2.33}$$

It follows from (2.2.32) that there exists a weakly convergent subsequence $u_\delta \rightharpoonup u$ as $\delta \to 0$, where we have denoted this subsequence also u_δ. Thus

$$||u|| \leq \liminf_{\delta\to 0} ||u_\delta||. \tag{2.2.34}$$

From (2.2.33) and (2.2.34) it follows that $||u|| \leq ||y||$. Let us prove that $Au = f$. Since $||u|| \leq ||y||$ and the minimal-norm solution to equation (2.2.1) is unique, we conclude that $u = y$. To verify the equation $Au = y$ we argue as follows: from (2.2.25) it follows that $\lim_{\delta\to 0} ||Au_\delta - f|| = 0$, so, for any $g \in D(A^*)$, one gets

$$(f, g) = \lim_{\delta\to 0}(Au_\delta, g) = \lim_{\delta\to 0}(u_\delta, A^*g) = (u, A^*g). \tag{2.2.35}$$

Since $A^{**} = \overline{A} = A$, where the overbar denotes the closure of A, this implies

$$(f - Au, g) = 0 \quad \forall g \in D(A^*).$$

Since $D(A^*)$ is dense, it follows that $Au = f$, as claimed. The density of $D(A^*)$ follows from the assumption that A is closed and densely defined.

Let us now prove (2.2.7). We have already proved that $u_\delta \rightharpoonup y$ and $||u_\delta|| \leq ||y||$. This implies (2.2.7). Indeed

$$||u_\delta - y||^2 = ||u_\delta||^2 + ||y||^2 - 2\text{Re}(u_\delta, y) \to 0 \text{ as } \delta \to 0.$$

Theorem 2.2.4 is proved. □

We have assumed in Theorem 2.2.4 that $u_{a,\delta} := T_a^{-1} A^* f_\delta$ is the exact minimizer of the functional (2.2.2). Suppose that $w_{a,\delta}$ is an approximate minimizer of (2.2.2) in the following sense

$$\mathcal{F}(w_{a,\delta}) \leq m + (c^2 - 1 - b)\delta^2, \quad c^2 > 1 + b, \tag{2.2.36}$$

where $c \in (1, 2)$ is a constant, $b > 0$ is a constant, and $m := \inf_u \mathcal{F}(u)$.

The problem is:

Will the equation

$$||Aw_{a,\delta} - f_\delta|| = c\delta, \tag{2.2.37}$$

be solvable for a for any $\delta > 0$ sufficiently small?

Will the element $w_\delta := w_{a(\delta),\delta}$ converge to y?

Our result is stated in the following new discrepancy principle.

Theorem 2.2.5. *Assume that (2.2.36) holds,*

$$||f_\delta|| > c\delta, \quad c = const \in (1, 2), \quad ||f_\delta - f|| \leq \delta, \quad f = Ay, \quad y \perp \mathcal{N}.$$

Then, for any $w_{a,\delta}$, satisfying (2.2.36) and depending continuously on a, equation (2.2.37) has a solution $a = a(\delta)$, such that $w_\delta = w_{a(\delta),\delta}$ converges to y:

$$\lim_{\delta \to 0} ||w_\delta - y|| = 0. \tag{2.2.38}$$

Proof of Theorem 2.2.5. Let $H(\delta, a) := ||Aw_{a,\delta} - f_\delta||$. Then $H(\delta, a)$ is continuous with respect to a because $w_{a,\delta}$ is continuous with respect to a. Let us verify that

$$H(\delta, +0) < c\delta, \quad H(\delta, \infty) > c\delta. \tag{2.2.39}$$

Then there exists $a = a(\delta)$ which solves (2.2.37).

As $a \to \infty$, we have

$$a||w_{a,\delta}||^2 \le \mathcal{F}(w_{a,\delta}) \le m + (c-1-b)\delta^2 \le \mathcal{F}(0) + (c^2-1-b)\delta^2. \quad (2.2.40)$$

Since $\mathcal{F}(0) = ||f_\delta||^2$, one gets from (2.2.40):

$$||w_{a,\delta}|| \le \frac{c}{\sqrt{a}}, \quad a \to \infty. \quad (2.2.41)$$

Therefore

$$\lim_{a\to\infty} ||w_{a,\delta}|| = 0,$$

so

$$H(\delta, \infty) = ||A0 - f_\delta|| = ||f_\delta|| > c\delta. \quad (2.2.42)$$

Let $a \to 0$. Then

$$H^2(\delta, a) \le \mathcal{F}(w_{a,\delta}) \le m + (c^2 - 1 - b)\delta^2 \le \mathcal{F}(y) + (c^2 - 1 - b)\delta^2.$$

One has

$$\mathcal{F}(y) = \delta^2 + a||y||^2.$$

So

$$H^2(\delta, a) \le (c^2 - b)\delta^2 + a||y||^2.$$

Thus

$$H(\delta, +0) \le (c^2 - b)^{\frac{1}{2}}\delta < c\delta. \quad (2.2.43)$$

From (2.2.43) and (2.2.42) one gets (2.2.39). So, the existence of the solution $a = a(\delta)$ of equation (2.2.37) is proved.

Let us prove (2.2.38). One has

$$\mathcal{F}(w_\delta) = ||Aw_\delta - f_\delta||^2 + a(\delta)||w_\delta||^2 = c^2\delta^2 + a(\delta)||w_{a,\delta}||^2 \le m + (c^2-1-b)\delta^2.$$

Thus, using the inequality $m \le \delta^2 + a(\delta)||y||^2$, one gets

$$c^2\delta^2 + a(\delta)||w_\delta||^2 \le a(\delta)||y||^2 + (c^2 - b)\delta^2.$$

Therefore

$$||w_\delta|| \le ||y||. \quad (2.2.44)$$

As in the proof of Theorem 2.2.4, inequality (2.2.44) and equation (2.2.37) imply $w_\delta \rightharpoonup w$, $Aw_\delta \longrightarrow f$ as $\delta \to 0$, and $Aw = f$. Since $||w|| \le ||y||$, as follows from (2.2.44), it follows that $w = y$, because y is the unique minimal-norm solution to equation (2.2.1). Thus, $w_\delta \rightharpoonup y$ and $||w_\delta|| \le ||y||$. This implies (2.2.38), as was shown in the proof of Theorem 2.2.4. Theorem 2.2.5 is proved. □

Let us discuss variational regularization for nonlinear equations. Let A be a possibly nonlinear, injective, closed map from a Banach space X into a Banach space Y. Let

$$F(u) := ||A(u) - f_\delta|| + \delta g(u),$$

where $g(u) \ge 0$ is a functional. Assume that the set $\{u : g(u) \le c\}$ is precompact in X. Assume that

$$A(y) = f, \quad ||f_\delta - f|| \le \delta, \quad \text{and} \quad D(A) \subset D(g),$$

so that $y \in D(g)$.

Theorem 2.2.6. *Under the above assumptions let u_δ be any sequence such that $F(u_\delta) \le c\delta$, $c := 2 + g(y)$. Then $\lim_{\delta\to 0} ||u_\delta - y|| = 0$.*

Proof of Theorem 2.2.6. Let

$$m := \inf_{u \in D(A)} F(u), \quad F(u_n) \le m + \frac{1}{n},$$

where $n = n(\delta)$ is the smallest positive integer satisfying the inequality $\frac{1}{n} \le \delta$. One has

$$m \le F(y) = \delta[1 + g(y)],$$

and

$$F(u_n) \le c\delta, \quad c = 2 + g(y).$$

Thus, $g(u_n) \le c$. Consequently, one can select a convergent subsequence u_δ, $||u_\delta - u|| \to 0$ as $\delta \to 0$. One has

$$0 = \lim_{\delta\to 0} F(u_\delta) = \lim_{\delta\to 0}\{||A(u_\delta) - f_\delta|| + \delta g(u_\delta)\} = \lim_{\delta\to 0} ||A(u_\delta) - f||.$$

Thus $u_\delta \to u$ and $A(u_\delta) \to f$. Since A is closed, this implies $A(u) = f$. Since A is injective and $A(y) = f$, one gets $u = y$. Thus

$$\lim_{\delta\to 0} ||u_\delta - y|| = 0.$$

Theorem 2.2.6 is proved.

Let us prove

Theorem 2.2.7. *Functional (2.2.2) has a unique global minimizer*

$$u_{a,\delta} = A^*(Q+a)^{-1} f_\delta$$

for any $f \in H$, where $Q := AA^*$, $a = const > 0$.

Proof of Theorem 2.2.7. Consider the equation

$$(Q+a)w_{a,\delta} = (AA^* + a)w_{a,\delta} = f_\delta.$$

It is uniquely solvable:

$$w_{a,\delta} = (Q+a)^{-1} f_\delta.$$

Define $u_{a,\delta} := A^* w_{a,\delta}$. Then

$$Au_{a,\delta} - f_\delta = -aw_{a,\delta}.$$

One has:

$$\begin{aligned} F(u+v) &= ||Au - f_\delta||^2 + a||u||^2 + ||Av||^2 + a||v||^2 \\ &+ 2\text{Re}[(Au - f_\delta, Av) + a(u,v)], \quad \forall v \in D(A). \end{aligned}$$

Let $u = u_{a,\delta}$. Then

$$(Au_{a,\delta} - f_\delta, Av) + a(u_{a,\delta}, v) = -a(w_{a,\delta}, Av) + a(u_{a,\delta}, v) = 0,$$

because

$$(w_{a,\delta}, Av) = (A^* w_{a,\delta}, v) = (u_{a,\delta}, v).$$

Consequently,

$$F(u_{a,\delta} + v) = F(u_{a,\delta}) + a||v||^2 + ||Av||^2 \geq F(u_{a,\delta}),$$

and $F(u_{a,\delta} + v) = F(u_{a,\delta})$ if and only if $v = 0$. Therefore

$$u_{a,\delta} = A^*(Q+a)^{-1} f_\delta$$

is the unique global minimizer of $F(u)$.

Theorem 2.2.7 is proved. □

Let us show how the results can be extended to the case when not only f is known with an error δ, but the bounded operator A is also known with an error δ, i.e., A_δ is known, $||A_\delta - A|| \leq \delta$, and A is unknown.

Consider, for instance, an analog of Theorem 2.2.1. Let us define

$$u_\delta := T^{-1}_{a(\delta),\delta} A^*_\delta f_\delta, \tag{2.2.45}$$

where

$$T_{a,\delta} := A_\delta^* A_\delta + aI. \tag{2.2.46}$$

The element

$$u_{a,\delta} := T_{a,\delta}^{-1} A_\delta^* f_\delta \tag{2.2.47}$$

solves the equation

$$(T_\delta + aI)u_{a,\delta} = A_\delta^* f_\delta. \tag{2.2.48}$$

We have

$$||T_\delta - T|| \leq ||(A_\delta^* - A^*)A_\delta|| + ||A^*(A_\delta - A)|| \leq 2\delta(||A|| + \delta). \tag{2.2.49}$$

Equation (2.2.48) can be written as

$$u_{a,\delta} = T_a^{-1}A^* + T_a^{-1}(A_\delta^* - A^*)f_\delta + T_a^{-1}A^*(f_\delta - f) + T_a^{-1}(A^*A - A_\delta^* A_\delta)u_{a,\delta}. \tag{2.2.50}$$

We have

$$||T_a^{-1}(A^*A - A_\delta^* A_\delta)|| \leq \frac{2\delta(||A|| + \delta)}{a}. \tag{2.2.51}$$

Assume that $a(\delta)$ satisfies the conditions

$$\lim_{\delta \to 0} \frac{\delta}{a(\delta)} = 0, \quad \lim_{\delta \to 0} a(\delta) = 0. \tag{2.2.52}$$

Then (2.2.51) implies that equation (2.2.50) is uniquely solvable for $u_{a(\delta),\delta} := u_\delta$, and

$$||u_\delta - T_{a(\delta)}^{-1} A^* f|| \leq c \left(\frac{\delta}{a(\delta)} ||f|| + \frac{\delta}{2\sqrt{a(\delta)}} \right) \underset{\delta \to 0}{\longrightarrow} 0, \tag{2.2.53}$$

where c is an upper bound of the norm of the operator $[I - (A^*A - A_\delta^* A_\delta)]^{-1}$. Thus $c = c(\delta)$, $\lim_{\delta \to 0} c(\delta) = 1$.

We have proved the following result.

Theorem 2.2.8. *Assume that f_δ and A_δ are given,*

$$||f_\delta - f|| \leq \delta, \quad ||A_\delta - A|| \leq \delta, \quad Ay = f, \quad y \perp \mathcal{N}(A),$$

u_δ is defined in (2.2.45), and (2.2.52) holds. Then $\lim_{\delta \to 0} ||u_\delta - y|| = 0$.

The discrepancy principle can also be generalized to the case when A_δ is given in place of A. This principle for choosing $a(\delta)$ can be formulated as the equation

$$||A_\delta T_{a,\delta}^{-1} A_\delta^* f_\delta - f_\delta|| = c\delta, \quad c \in (R, R+1), \tag{2.2.54}$$

where $R \geq 1 + ||y||$ is a constant, and we assume that

$$||f_\delta|| > c\delta. \tag{2.2.55}$$

Equation (2.2.54) is uniquely solvable for $a = a(\delta)$, and $\lim_{\delta\to 0} = 0$. This is proved as in the proof of Theorem 2.2.4. Condition (2.2.54) allows one to get the estimate for $u_\delta := u_{a(\delta),\delta}$, where $a(\delta)$ solves (2.2.54). This estimate

$$||u_\delta|| \leq ||y|| \tag{2.2.56}$$

is similar to the estimate (2.2.32). It allows one to derive the relation (2.2.7). The constant $1 + ||y||$ is a lowere bound for R in (2.2.54) because the inequality

$$||A_\delta u_{a,\delta} - f_\delta||^2 + a||u_{a,\delta}|| \leq ||A_\delta y - f_\delta||^2 + a||y||^2 \tag{2.2.57}$$

and the relation (2.2.54) imply inequality (2.2.56), provided that

$$c \geq ||y|| + 1. \tag{2.2.58}$$

This follows from the estimate:

$$\begin{aligned} ||A_\delta y - f_\delta|| &\leq ||(A_\delta - A)y|| + ||Ay - f_\delta|| \\ &\leq \delta||y|| + \delta = \delta(||y|| + 1). \end{aligned} \tag{2.2.59}$$

Thus, we have proved the following result similar to Theorem 2.2.4.

Theorem 2.2.9. *Equation (2.2.54) is uniquely solvable for $a = a(\delta)$ provided that (2.2.25)holds. The relation $\lim_{\delta\to 0} a(\delta) = 0$ holds. The element $u_\delta := u_{a(\delta),\delta}$ satisfies the relation (2.2.7) provided that (2.2.58) holds.*

Although the solution y is unknown, an upper bound on y is often known a priori as a part of a priori information about the unkown solution.

2.3 Quasisolutions

Let us assume that $A : X \to Y$ is a continuous operator from a Banach space X into a Banach space Y, $K \subset X$ is a compact set.

Definition 2.3.1. *An element $z \in K$ is called a quasisolution of the equation $Au = f$ if*

$$||Az - f|| = \inf_{u \in K} ||Au - f||.$$

If A is continuous, then the functional $||Au - f||$ is continuous. Every continuous functional achieves on a compact set its infimum. Therefore quasisolutions are well defined for any $f \in Y$.

Under suitable assumptions one can prove that the quasisolution not only exists, but is unique and depends continuously on f. To formulate these assumptions, let us recall some geometrical notions.

Definition 2.3.2. *A Banach space is called strictly convex if $||u + v|| = ||u|| + ||v||$ implies $u = \lambda v$, $\lambda =$ const $\in \mathbb{R}$.*

Definition 2.3.3. *Let $M \subset X$ be a convex set, i.e. $u, v \in M$ implies $\lambda u + (1 - \lambda)v \in M$ for all $\lambda \in (0, 1)$. Then metric projection of an element $w \in X$ onto M is an element $Pw = v \in M$, such that*

$$||w - v|| = \inf_{z \in M} ||w - z||.$$

Lemma 2.3.1. *In a strictly convex Banach space X the metric projection onto a convex set M is unique.*

Proof of Lemma 2.3.1 Suppose that v_1 and v_2 are metric projections of $u \notin M$ onto M, so that

$$||u - v_1|| = ||u - v_2|| = \inf_{z \in M} ||u - z|| := m > 0.$$

Then, since M is convex, $\frac{v_1 + v_2}{2} \in M$, and we get:

$$m \leq ||u - \frac{v_1 + v_2}{2}|| \leq \frac{1}{2}(||u - v_1|| + ||u - v_2||) = m.$$

Thus

$$||u - v_1|| = ||u - v_2|| = ||\frac{u - v_1 + u - v_2}{2}|| > 0.$$

Denote $u - v_1 = p$, $u - v_2 = q$. Then $||p|| = ||q|| = \frac{1}{2}||p + q|| > 0$, so $||p + q|| = ||p|| + ||q||$. Since X is strictly convex, it follows that $p = \lambda q$, $|\lambda| = 1$, so $\lambda = 1$ or $\lambda = -1$, because λ is real-valued. If $\lambda = -1$ then $||p + q|| = 0$, a contradiction. So $\lambda = 1$. Thus $p = q$, so $v_1 = v_2$. Lemma 2.3.1 is proved. □

Remark 2.3.1. Hilbert space is strictly convex, Lebesgue's spaces $L^p(D)$, $1 < p < \infty$, are strictly convex, but $C(D)$ and $L^1(D)$ are not strictly convex.

Lemma 2.3.2. *The operator P of metric projection onto a convex compact set M of a strictly convex Banach space is continuous.*

Proof of Lemma 2.3.2. We claim that the distance

$$d(f, M) = \inf_{z \in M} ||f - z||$$

is a continuous function of f for any set M, not necessarily convex or compact.

Indeed,

$$d(f_1, M) \leq d(f_1, z) \leq d(f_1, f_2) + d(f_2, z),$$

so

$$d(f_1, M) - d(f_2, M) \leq d(f_1, f_2).$$

By symmetry,

$$d(f_2, M) - d(f_1, M) \leq d(f_1, f_2).$$

Thus

$$|d(f_1, M) - d(f_2, M)| \leq d(f_1, f_2),$$

as claimed.

Let us prove the continuity of $P = P_M$. Let $\lim_{n\to\infty} ||f_n - f|| = 0$, and assume that $||q_n - q|| > \varepsilon > 0$, where $q_n = P_M f_n$ and $q = P_M f$. Since M is compact and q is bounded (by the above claim), one may select a convergent subsequence, which we denote again by q_n,

$$q_n \to v \in M, \quad ||v - q|| \geq \varepsilon > 0.$$

One has $||f - q|| \leq ||f - v||$ and

$$||f - v|| \leq ||f - f_n|| + ||f_n - q_n|| + ||q_n - v||.$$

Note that

$$\lim_{n\to\infty} ||f - f_n|| = 0, \quad \lim_{n\to\infty} ||q_n - v|| = 0,$$

and, by the claim,

$$\lim_{n\to\infty} ||f_n - q_n|| = \lim_{n\to\infty} d(f_n, M) = d(f, M) = ||f - q||.$$

Therefore

$$||f - q|| = ||f - v||.$$

Lemma 2.3.2 is proved. □

From Lemmas 2.3.1 and 2.3.2 the following result follows.

Theorem 2.3.1. *Assume that A is a linear continuous injection, $M \subset X$ is convex and compact, and X is strictly convex. Then the quasisolution to equation $A(u) = f$ exists, is unique, and depends continuously on f.*

Proof of Theorem 2.3.1. Existence of the quasisolution follows from compactness of M and continuity of A, as was explained below Definition 2.3.1. Uniqueness of the quasisolution follows from the injectivity of A. The quasisolution depends continuously on f by Lemma 2.3.2 because the set AM is convex and compact, so the map $f \to P_{AM}f$ is continuous, and the quasisolution $u = A^{-1}P_{AM}f$ is continuous on the set AM if M is compact.

The continuity of A^{-1} on AM is a consequence of the following lemma.

Lemma 2.3.3. *Assume that $A : M \to X$ is a possibly nonlinear, injective, closed map from a compact set M of a complete metric space X into X. Then the inverse map A^{-1} is continuous on AM.*

Proof of Lemma 2.3.3. Let $A(u_n) = f_n$, $f_n \to f$ as $n \to \infty$, $u_n \in M$. Since M is compact, one can select a convergent subsequence, which we denote u_n again, $u_n \to u \in M$. Since A is closed, the convergence $u_n \to u$, $A(u_n) \to f$ implies $A(u) = f$. Since A is injective, the equation $A(u) = f$ defines uniquely $u = u_f = A^{-1}(f)$. Thus $A^{-1}(f_n) \to A^{-1}(f)$. Since the limit of every subsequence u_n is the same, $A^{-1}(f)$, the sequence $u_n = A^{-1}(f_n)$ converges to $A^{-1}(f)$. Lemma 2.3.3 is proved. □

Remark 2.3.2. Lemma 2.3.3 differs from the well-known result [[DS], Lemma I.5.8] because the continuity of A is replaced by the closedness of A.

Let us now consider quasisolutions for nonlinear equations. Let $A : X \to Y$ be an injective, possibly nonlinear, closed map from a Banach space X into a Banach space Y. Let $K \subset D(A) \subseteq X$ be a compact set. Assume that

$$A(y) = f \quad \text{and} \quad ||f_\delta - f|| \leq \delta.$$

Let

$$m := \inf_{u \in K} ||A(u) - f_\delta||.$$

Choose a minimizing sequence u_n such that

$$||A(u_n) - f - \delta|| \leq m + \frac{1}{n}, \quad n = 1, 2.....$$

Let $n = n(\delta)$ be the smallest positive integer such that $\frac{1}{n} \leq \delta$, so that

$$||A(u_n) - f_\delta|| \leq m + \delta.$$

Since

$$m \leq ||A(y) - f_\delta|| = \delta,$$

one has

$$||A(u_n) - f_\delta|| \leq 2\delta.$$

Thus

$$||A(u_n) - f|| \to 0 \quad \text{as} \quad \delta \to 0.$$

Since $u \in K$ and K is compact, one can select a convergent subsequence $u_{\delta_j} := v_j$, $v_j \to v$ as $j \to \infty$, $v \in K$. Thus $A(v_j) \to f$, $v_j \to v$, so $A(v) = f$, because A is closed. Since A is injective, the element v is uniquely determined by f. Therefore $v = y$, and

$$\lim_{\delta \to 0} ||u_\delta - y|| = 0.$$

We have proved the following theorem.

Theorem 2.3.2. *Let $A : X \to Y$ be an injective, closed, possibly nonlinear, map from a Banach space X into a Banach space Y. Let $K \subset D(A)$ be a compact set. Assume that $A(y) = f$, and $||f_\delta - f|| \leq \delta$. Let $u_\delta \in K$ be any sequence such that $||A(u_\delta) - f_\delta|| \leq 2\delta$. Then $\lim_{\delta \to 0} ||u_\delta - y|| = 0$.*

2.4 Iterative regularization

The main result in this Section is the following theorem.

Theorem 2.4.1. *Every solvable linear equation $Au = f$ with a closed densely defined operator A in a Hilbert space H can be solved by a convergent iterative process.*

Proof of Theorem 2.4.1. Let $T = A^*A, \quad u, \quad Q = AA^*$ be nonnegative selfadjoint operators, $a = \text{const}$, $T_a = T + aI$, I is the identity operator, $B = aT_a^{-1}$ we have proved (see Section 2.2, proof of Theorem 2.2.2) that $T_a^{-1}A^*$ is a bounded operator defined on all of H, $||T_a^{-1}A^*|| \leq \frac{1}{2\sqrt{a}}$. Consider an iterative process

$$u_{n+1} = Bu_n + T_a^{-1}A^*f, \quad u_1 \perp \mathcal{N}, \quad B = aT_a^{-1}, \tag{2.4.1}$$

where $\mathcal{N} = \mathcal{N}(A) = \mathcal{N}(T)$. Denote

$$w_n := u_n - y,$$

where $Ay = f$, $y \perp \mathcal{N}$. One has

$$y = By + T_a^{-1}A^*f. \tag{2.4.2}$$

Thus

$$w_{n+1} = Bw_n, \quad w_1 = u_1 - y \perp \mathcal{N}. \tag{2.4.3}$$

Therefore,

$$w_{n+1} = B^n w_1, \quad w_1 \perp \mathcal{N}. \tag{2.4.4}$$

Let E_s be the resolution of the identity corresponding to the selfadjoint operator T. Then

$$\begin{aligned} \lim_{n\to\infty} ||w_{n+1}||^2 &= \lim_{n\to\infty} \int_0^\infty \frac{a^{2n}}{(a+s)^{2n}} d(E_s w_1, w_1) \\ &= ||(E_0 - E_{-0})w_1||^2 = ||P_N w_1||^2 = 0, \end{aligned} \tag{2.4.5}$$

where P_N is the orthoprojector onto $\mathcal{N}$.

Remark 2.4.1. If A is a bounded operator, then the operator T_a^{-1} can be easily computed if $a > ||T|| = ||A||^2$. Indeed

$$T_a^{-1} = (T + aI)^{-1} = a(I + a^{-1}T)^{-1},$$

and $||T||a^{-1} < 1$ if $a > ||T||$, so that

$$(I + a^{-1}T)^{-1} = \sum_{j=0}^{\infty} (-1)^j a^{-j} T^j,$$

and the series converges at the rate of geometrical series.

Remark 2.4.2. One cannot give a rate of convergence in (2.4.5) without extra assumptions on w_1. If, for example, $w_1 = T$, then the integral in (2.4.5) can be written as

$$\int_0^\infty \frac{a^{2n}s^2}{(a+s)^{2n}} d(E_s z, z).$$

One can check that

$$\max_{s\geq 0} \frac{a^{2n}s^2}{(a+s)^{2n}} \leq c\frac{a^2}{n^2},$$

where $c = \text{const} > 0$ does not depend on a and n,

$$c = \max \frac{n^2}{(n-1)^2} \frac{1}{(1+\frac{1}{n-1})^{2n}} \leq 4.$$

Thus, if the extra assumption on w_1 is $w_1 = Tz$, then the rate of decay in (2.4.5) is

$$||w_{n+1}||^2 \leq c\frac{a^2}{n^2}||z||^2.$$

A similar calculation can be made if $w_1 = T^b z$, where $b > 0$ is a constant.

Remark 2.4.3. An idea, similar to the one used in the proof of Theorem 2.4.1, can be applied to the equation

$$Au = f, \tag{2.4.6}$$

where $A = A^*$ is selfadjoint not necessarily bounded operator, and we assume that

$$Ay - f, \quad y \perp \mathcal{N}.$$

Let $a > 0$ be a constant. Equation (2.2.6) is equivalent to

$$u = iaA_{ia}^{-1}u + A_{ia}^{-1}f, \quad A_{ia} := A + iaI. \tag{2.4.7}$$

Consider the iterative process

$$u_{n+1} = iaA_{ia}^{-1}u_n + A_{ia}^{-1}f, \quad u_1 \perp \mathcal{N}. \tag{2.4.8}$$

Theorem 2.4.2. *If* $A = A^*$ *and* $a > 0$, *then* $lim_{n\to\infty}||u_n - y|| = 0$, *provided that* $u_1 \perp \mathcal{N}$.

Proof of Theorem 2.4.2. Let $v_n := u_n - y$. Since

$$y = iaA_{ia}^{-1}y + A_{ia}^{-1}f,$$

one gets

$$v_{n+1} = iaA_{ia}^{-1}v_n, \quad v_1 = u_1 - y \perp \mathcal{N}. \tag{2.4.9}$$

Thus $v_{n+1} = (ia)^n A_{ia}^{-1} v_1$, so

$$\lim_{n\to\infty} ||v_{n+1}||^2 = \lim_{n\to\infty} \int_{-\infty}^{\infty} \frac{a^{2n} d(E_s v_1, v_1))}{(a^2+s^2)^n} = ||P_{\mathcal{N}} v_1||^2 = 0, \tag{2.4.10}$$

where E is the resolution of the identity, corresponding to A, and $\mathcal{N} = \mathcal{N}(A)$.

Theorem 2.4.2 is proved. □

Let us discuss another iterative process for solving equation $Au = f$. We assume that this equation is solvable, and $y \perp \mathcal{N}$ is its minimal-norm solution. Let us also assume that A is bounded. Let $T = A^*A$. If $Au = f$, then $Tu = A^*f$. Conversely, if equation $Au = f$ is solvable, then every solution to the equation $Tu = A^*f$ solves equation $Au = f$. Indeed, write $f = Ay$. Then $Tu = Ty$. Multiply this equation by $u - y$ and get $A(u - y) = 0$. Thus $Au = Ay = f$.

The iterative process for solving linear solvable equation $Au = f$ can be written as follows:

$$u_{n+1} = u_n - (Tu_n - A^*f), \quad u_1 = 0, \tag{2.4.11}$$

provided that

$$||T|| \le 1. \tag{2.4.12}$$

Note that (2.4.12) is not really a restriction because one can always divide the equation $Tu = A^*f$ by $||T|| > 0$ and denote $\frac{T}{||T||}$ by T_1. Then $||T_1|| = 1$, so T_1 satisfies the restriction (2.4.12).

Theorem 2.4.3. *One has* $\lim_{n\to\infty} ||u_n - y|| = 0$, *where* $Ay = f$, $y \perp \mathcal{N}$, *(2.4.12) is assumed, and* u_n *is defined by (2.4.11).*

Proof of Theorem 2.4.3. Let $u_n - y := z_n$. One has

$$y = y - (Ty - A^*f).$$

Thus

$$z_{n+1} = z_n - Tz_n = (I - T)^n z_1. \tag{2.4.13}$$

Let E_s be the resolution of the identity corresponding to T. Then

$$\lim_{n\to\infty} ||z_{n+1}||^2 = \lim_{n\to\infty} \int_0^1 (1-s)^{2n} d(E_s z_1, z_1) = ||P_N z_1||^2 = 0, \quad (2.4.14)$$

because $z_1 = u_1 - y = -y \perp \mathcal{N}$.

Theorem 2.4.3 is proved. □

2.5 Quasiinversion

The quasiinversion method (see [LL]) was applied to the ill-posed problem for the heat equation considered in Example 2.1.20. Following [LL], we consider a more general setting. Let

$$\dot{u} - Au = 0, \quad u(0) = f, \quad (2.5.1)$$

where $A = A^* \geq c > 0$ is a selfadjoint operator in a Hilbert space H. The problem is:

Given $g \in H$, a number $T > 0$, and an arbitrary small $\varepsilon > 0$, find f such that $||u(T) - g|| \leq \varepsilon$.

The solution $u(t) = u(t; f)$ to (2.5.1) exists and is unique. It can be written as

$$u = e^{tA} f = \int_c^\infty e^{-ts} dE_s f,$$

where E_s is the resolution of the identity for A.

Note that if there is an f such that $u(T; f) = g$, then such an f is unique. Indeed, if there are two such f, f_1 and f_2, then $0 = \int_c^\infty e^{-Ts} dE_s f$, where $f := f_1 - f_2$, so

$$0 = \int_c^\infty e^{-Ts} d(E_s f, f), \qquad c > 0.$$

Since $(E_s f, f)$ is a monotonically nondecreasing function, it follows that $f = 0$.

It is also easy to check that, for any fixed $g \in H$, one has

$$\inf_{f \in H} ||u(T; f) - g|| = 0. \quad (2.5.2)$$

Indeed, otherwise one would have a nonzero element g_1 orthogonal to the span of $u(T; f)$ for all $f \in H$, i.e. $(u(T; f), g_1) = 0 \quad \forall f \in H$. Taking $f = g_1$, one gets

$$\int_c^\infty e^{-Ts} d(E_s g_1, g_1) = 0.$$

This implies $g_1 = 0$. Thus, (2.5.2) is proved.

The problem

$$\dot{u} + Au = 0, \quad u(T) = g, \tag{2.5.3}$$

can be reduced to the problem

$$-\frac{du}{d\tau} + Au = 0, \quad u(0) = g, \quad \tau = T - t. \tag{2.5.4}$$

Consider the set

$$M := \{u : ||u(T)|| \le c\},$$

where u solves (2.5.4).

Lemma 2.5.1. *If $u \in M$ and $||g|| \le \varepsilon$, then*

$$||u(\tau)|| \le c^{\frac{\tau}{T}} \varepsilon^{\frac{T-\tau}{T}}, \quad 0 \le \tau \le T. \tag{2.5.5}$$

Proof of Lemma 2.5.1. Let $h := ||u(\tau)||$. Then

$$\dot{h} := \frac{dh}{d\tau} = 2\text{Re}(\dot{u}, u), \quad \ddot{h} = 2||\dot{u}||^2 + 2\text{Re}(\ddot{u}, u).$$

One has

$$(\ddot{u}, u) = (Au, Au) = ||\dot{u}||^2,$$

so

$$\ddot{h} = 4||\dot{u}||^2.$$

Define $p(\tau) := \ln h$. Then

$$\ddot{p} = \frac{\ddot{h}h - \dot{h}^2}{h^2} = \frac{4||\dot{u}||^2 h - 4|Re(\dot{u}, u)|^2}{h^2} \ge 0.$$

Therefore $p(\tau)$ is a convex function. Thus

$$p(\tau) \le \frac{T - \tau}{T} p(0) + \frac{\tau}{T} p(T).$$

This implies

$$h(\tau) \le [h(0)]^{\frac{T-\tau}{T}} [h(T)]^{\frac{\tau}{T}},$$

or

$$||u(\tau)|| \leq c^{\frac{\tau}{T}} \varepsilon^{\frac{T-\tau}{T}}.$$

Lemma 2.5.1 is proved. □

Estimate (2.5.5) gives a continuous dependence of the solution to (2.5.4) or (2.5.3) on g under a priori assumption $u \in M$.

Given g, the quasiinversion method for finding f in the sense

$$||g - e^{-TA}f|| \leq \eta, \tag{2.5.6}$$

where $\eta > 0$ is an arbitrary small given number, consists of solving the problem

$$\dot{u_\varepsilon} + Au_\varepsilon - \varepsilon A^2 u_\varepsilon = 0, \quad u(T) = g, \quad t \geq T, \tag{2.5.7}$$

and then finding $f = f_\eta = u_\varepsilon(0)$. Here $\varepsilon = \text{const} > 0$. If $\varepsilon > 0$ is sufficiently small, then

$$||g - e^{-TA}u_\varepsilon(0)|| \leq \eta.$$

Let us justify the above claim. The unique solution to (2.5.7) is:

$$u_\varepsilon(t) = e^{-(t-T)(A-\varepsilon A^2)}g, \tag{2.5.8}$$

where the element $u_\varepsilon(t)$,

$$u_\varepsilon(t) = e^{-(t-T)(A-\varepsilon A^2)}g = \int_c^\infty e^{-(t-T)(s-\varepsilon s^2)}dE_s g,$$

is well defined for any $g \in H$ and for $t \in [0, T)$ provided that $\varepsilon > 0$. Let

$$f = f_\varepsilon := u_\varepsilon(0) = e^{T(A-\varepsilon A^2)}g. \tag{2.5.9}$$

Then

$$||g - e^{-TA}f_\varepsilon||^2 = ||g - e^{-\varepsilon TA^2}g||^2 = \int_c^\infty |1 - e^{-\varepsilon Ts^2}|^2 d(E_s g, g). \tag{2.5.10}$$

Thus, for any fixed $g \in H$, one can pass to the limit $\varepsilon \to 0$ in (2.5.10) and get

$$\lim_{\varepsilon \to 0} ||g - e^{-TA}f_\varepsilon|| = 0. \tag{2.5.11}$$

Therefore, for any $g \in H$ and any $\eta > 0$, however small, one can find f_ε such that

$$||g - e^{-TA} f_\varepsilon|| \le \eta.$$

If $g = e^{-TA}h$ for some $h \in H$, i.e., g is the value of a solution to equation (2.5.1) at $t = T$ with $u(0) = h$, then

$$||g - e^{-TA} f_\varepsilon|| \le ||e^{-TA}|| \, ||h - f_\varepsilon|| \le ||h - f_\varepsilon||.$$

Therefore, if $g = e^{-TA}h$, $h \in H$, then taking any f satisfying the inequality $||f_\varepsilon - h|| \le \eta$, implies $||g - e^{-TA} f_\varepsilon|| \le \eta$.

Remark 2.5.1. In [LL] the case when $A = A(t)$ is treated under suitable assumptions.

Remark 2.5.2. There are infinitely many elements f satisfying (2.5.6). Formula (2.5.9) gives one concrete such an element.

Remark 2.5.3. The quasiinversion method yields stable solution of the above problem: if g_δ is given, $||g_\delta - g|| \le \delta$, then

$$||e^{T(A-\varepsilon A^2)}(g_\delta - g)|| \le ||e^{T(A-\varepsilon A^2)}||\delta = N(\varepsilon, T)\delta, \tag{2.5.12}$$

where

$$N(\varepsilon, T) = \sup_{s \ge c} e^{T(s-\varepsilon s^2)} = e^{\frac{T}{4\varepsilon}}. \tag{2.5.13}$$

Remark 2.5.4. In [LL] the quasiinversion method is applied to solving Cauchy problems for elliptic equations and to other ill-posed problems.

2.6 Dynamical systems method (DSM)

The idea of a DSM has been described in Section 1.1.

We want to solve an operator equation

$$F(u) - f = 0 \tag{2.6.1}$$

in a Hilbert space H. We assume that this equation has a solution y, that

$$\sup_{u \in B(u_0, R)} ||F^{(j)}(u)|| \le M_j(R), \quad j = 0, 1, 2, \tag{2.6.2}$$

where $F^{(j)}(u)$ are Fréchet derivatives of F, $u_0 \in H$ is some element,

$$B(u_0, R) = \{u : ||u - u_0|| \le R, \quad u \in H\},$$

and $M_j(R)$ are some positive constants. We do not restrict the growth of these constants as $R \to \infty$. This means that the nonlinearity F can grow arbitrarily fast as $||u||$ grows, but is locally twice Fréchet differentiable.

Definition 2.6.1. *We will call problem (2.6.1) well-posed (WP) if*

$$\sup_{u\in B(u_0,R)} ||[F'(u)]^{-1}|| \leq m(R), \tag{2.6.3}$$

and ill-posed (IP) otherwise.

Condition (2.6.3) implies that F is a local homeomorphism in a neighborhood of the point u.

If $F(u) = Au$, where A is a bounded linear operator, then $F'(u) = A$ for any $u \in H$, and condition (2.6.3) implies that A is an isomorphism.

If $F(u) = Au$ so that $F'(u) = A$, and A is not boundedly invertible, i.e. (2.6.3) fails, then either A is not injective, or A is not surjective, or A^{-1} is not bounded. If A is not injective, i.e. $\mathcal{N} = \mathcal{N}(A) \neq \{0\}$, then one can consider A as a mapping from the factor space $H/\mathcal{N}$ into H, and then this mapping is injective, or one can consider A as a mapping from $H_1 = H \ominus \mathcal{N}$, and this mapping is injective.

If A is not surjective but its range $\mathcal{R}(A)$ is closed, then A as a mapping from H_1 into $\mathcal{R}(A)$ is injective, surjective, and has a bounded inverse, so in this case the problem of solving equation $Au - f = 0$ may be reduced to a well-posed problem:

If $||f_\delta - f|| \leq \delta$, but $f \notin \mathcal{R}(A)$, then one projects f_δ onto $\mathcal{R}(A)$, and the solution $A^{-1}P_{\mathcal{R}(A)}f_\delta$ depends continuously on f_δ.

However, if the range $\mathcal{R}(A)$ is not closed, $\mathcal{R}(A) \neq \overline{\mathcal{R}(A)}$, then small perturbations f_δ, $||f_\delta - f|| \leq \delta$, of f may lead to large perturbations of the solution, or f_δ may be out of $\mathcal{R}(A)$, in which case the equation $Au = f_\delta$ is not solvable. Thus, if $\mathcal{R}(A) \neq \overline{\mathcal{R}(A)}$, then the problem of solving equation $Au = f$ is ill-posed.

Calculating the null space $\mathcal{N}$ of a linear operator A is an ill-posed problem: a small perturbation of A may transform A into an injective operator.

If F is a nonlinear mapping, then condition (2.6.3), in general, does not imply injectivity or surjectivity globally. For example, if $F(u) = e^u$ is a mapping of $\mathbb{R}$ into $\mathbb{R}$, then

$$F'(u) = e^u, \quad |[F'(u)]^{-1}| = |e^{-u}| \leq m(R), \quad |u| \leq R,$$

however equation $e^u = 0$ has no solution. An example of noninjectivity: Let

$$F : \mathbb{R}^2 \to \mathbb{R}^2, \quad F(u) = \begin{pmatrix} e^{u_1}\cos u_2 \\ e^{u_1}\sin u_2 \end{pmatrix}.$$

Then

$$F'(u) = \begin{pmatrix} \frac{\partial F_1}{\partial u_1} & \frac{\partial F_1}{\partial u_2} \\ \frac{\partial F_2}{\partial u_1} & \frac{\partial F_2}{\partial u_2} \end{pmatrix} = \begin{pmatrix} e^{u_1}\cos u_2 & -e^{u_1}\sin u_2 \\ e^{u_1}\sin u_2 & e^{u_1}\cos u_2 \end{pmatrix},$$

$$[F'(u)]^{-1} = e^{-u_1} \begin{pmatrix} \cos u_2 & \sin u_2 \\ -\sin u_2 & \cos u_2 \end{pmatrix},$$

so

$$||[F'(u)]^{-1}|| \le e^{-u_1} \le e^R, \quad |u_1|^2 + |u_2|^2 \le R^2.$$

Condition (2.6.3) holds with any $R > 0$, but F is not injective: if equation $F(u) = f$ has a solution $\begin{pmatrix} u_1 \\ u_2 \end{pmatrix}$, then it has solutions $\begin{pmatrix} u_1 \\ u_2 + 2n\pi \end{pmatrix}$, $n = \pm 1, \pm 2, \ldots$.

The DSM method for solving equation (2.6.1), which consists of finding a nonlinear mapping $\Phi(t, u)$ such that the Cauchy problem

$$\dot{u} = \Phi(t, u), \quad u(0) = u_0, \tag{2.6.4}$$

has a unique global solution, there exists $u(\infty)$, and $u(\infty)$ solves equation (2.6.1):

$$\exists! u(t) \quad \forall t \ge 0; \quad \exists u(\infty); \quad F(u(\infty)) - f = 0. \tag{2.6.5}$$

If F is nonlinear, then the global existence of a solution to (2.6.4) is a difficult problem by itself. To guarantee the local existence and uniqueness of the solution to (2.6.4), we assume that Φ satisfies a Lipschitz condition

$$||\Phi(t, u) - \Phi(t, v)|| \le K||u - v||, \quad u, v \in B(u_0, R), \tag{2.6.6}$$

where $K = \text{constant} > 0$ does not depend on t, u, v but may depend on R. We also assume that

$$\sup_{t \in (t_0, t_0 + b)\ u \in B(u_0, R)} ||\Phi(t, u)|| \le K_0(R) := K_0, \tag{2.6.7}$$

where $K_0 > 0$ and $b > 0$ are a constants. Such an assumption on Φ will be satisfied in most of our applications. Assumption (2.6.6) does not restrict the growth of nonlinearity of Φ because K may grow rapidly as R grows.

Problem (2.6.4), in general, may have no global solution: the solution may blow up in a finite time. The maximal interval $[0, t)$ of the existence of the solution may be finite, $T < \infty$. In all the problems we deal with in this monograph, the global existence of the solution to (2.6.4) will be proved by establishing an a priori bound for the norm of the solution:

$$\sup_{t>0} ||u(t)|| \leq c, \tag{2.6.8}$$

where $c = \text{const} > 0$ does not depend on t and the supremum is taken over all $t \in [0, T)$.

Lemma 2.6.1. *If (2.6.6) and (2.6.8) hold, then $T = \infty$.*

Proof of Lemma 2.6.1. From the classical local existence and uniqueness theorem for the solution to (2.6.4) under the assumptions (2.6.6), (2.6.7) one obtains that the solution to (2.6.4) with the initial condition $u(t_0) = u_0$ exists and is unique on the interval $|t - t_0| < \tau$, where

$$\tau = \min\left(\frac{1}{K}, \frac{R}{K_0}, b\right). \tag{2.6.9}$$

This result is well-known, but we prove it for the convenience of the reader (see also Section 16.2). Using this result, we complete the proof of Lemma 2.6.1.

Assume that (2.6.8) holds but $T < \infty$. Let $u \in B(u_0, R)$ for $t \in [0, T - \frac{\tau}{2}]$. Take $t_0 = T - \frac{\tau}{2}$. Then the solution $u(t)$ to (2.6.4) with the initial condition $u(t_0) = u(T - \frac{\tau}{2})$ exists on the interval $(T - \frac{\tau}{2}, T + \frac{\tau}{2})$, so T is not the maximal interval of existence. Condition (2.6.8)guarantees that the Lipschitz constant K in (2.6.6) remains bounded, $K = K(c)$, the constant K_0 in (2.6.7) remains bounded, $K_0 = K_0(c)$, and the radius of the ball $B(u_0, R)$, within which the solution stays, remains bounded, $R = R(c)$. Thus, $\tau = \tau(c)$ does not decrease as the initial point gets close to T. This and the local existence and uniqueness theorem for the solution to (2.6.4) imply $T = \infty$. Lemma 2.6.1 is proved. □

In the following Theorem 2.6.1 we assume that $\Phi(t, u)$ is continuous operator function of t.

Theorem 2.6.1. *(local existence and uniqueness theorem).*

Assume (2.6.6)-(2.6.7). Then problem (2.6.4) with $u(t_0) = u_0$, $t_0 > 0$ has a unique solution $u(t) \in B(u_0, R)$ on the interval $t \in (t, t + \tau)$, where τ is defined in (2.6.9).

Proof of theorem 2.6.1. Problem (2.6.4) with $u(t_0) = u_0$, is equivalent to the equation:

$$u = u_0 + \int_{t_0}^{t} \Phi(s, u(s))ds := B(u), \tag{2.6.10}$$

because we have assumed that Φ is continuous with respect to t. Let us check that operator B maps the ball $B(u_0, R)$ into itself and is a contraction mapping on this ball if (2.6.9) holds. This, and the contraction mapping principle yield the conclusion of Theorem 2.6.1.

We have

$$\sup_{t\in(t_0,t_0+\tau)} ||u(t) - u_0|| \le \sup_{s\in(t_0,t_0+\tau),\ u\in B(u_0,R)} ||\Phi(s, u(s))|| \le \tau K_0 \le R. \tag{2.6.11}$$

Furthermore,

$$\begin{aligned}\sup_{t\in(t_0,t_0+\tau)} &||B(u) - B(v)|| \\ &\le \tau \sup_{s\in(t_0,t_0+\tau),\ u\in B(u_0,R)} ||\Phi(s, u(s)) - \Phi(s, v(s))|| \\ &\le \tau K \sup_{s\in(t_0,t_0+\tau)} ||u(s) - v(s)||.\end{aligned} \tag{2.6.12}$$

Since $\tau K > 1$, the operator B is a contraction on the set $B(u_0, R)$. Theorem 2.6.1 is proved. □

2.7 Variational regularization for nonlinear equations

Consider the equation

$$F(u) = f, \tag{2.7.1}$$

where F is a nonlinear map in a Hilbert space H.

Assume that there is a solution y to equation (2.7.1), $F(y) = f$.

We also assume that F is a weakly continuous (wc) map, i.e.

$$u_n \rightharpoonup u \quad \Rightarrow F(u_n) \rightharpoonup F(u). \tag{2.7.2}$$

Let $||f_\delta - f|| \le \delta$ and

$$g(u) = ||F(u) - f_\delta|| + b\delta\varphi(u), \quad b = \text{const} > 0, \tag{2.7.3}$$

where $\varphi : H_1 \to \mathbb{R}_+ := [0, \infty)$ is a functional, $H_1 \subset H$ is a Hilbert space, $||u||_1 \ge ||u||$, the embedding $i : H_1 \to H$ is compact, i.e. the set $\{u : ||u||_1 \le c\}$ contains a convergent in H subsequence, φ is weakly lower semicontinuous (wlsc), i.e.

$$u_n \rightharpoonup u \quad \Rightarrow \quad \underline{\lim}_{n\to\infty} \varphi(u_n) \ge \varphi(u), \tag{2.7.4}$$

and $y \in H_1$. From (2.7.2) and (2.7.4) it follows that g is wlsc.

Lemma 2.7.1. *If (2.7.2) and (2.7.4) hold, then the problem*

$$g(u) = \min \tag{2.7.5}$$

has a solution.

Proof. Since $g(u) \geq 0$, there exists $m := \inf_u g(u)$. Let $g(u_n) \to m$. Then $0 \leq g(u_n) \leq m + \delta$ for all $n \geq n(\delta)$, so

$$b\delta\varphi(u_n) \leq m + \delta. \tag{2.7.6}$$

We have

$$m \leq g(y) \leq \delta(1 + b\varphi(y)). \tag{2.7.7}$$

From (2.7.6) and (2.7.7) we get

$$\varphi(u_n) \leq \frac{2 + b\varphi(y)}{b} := c. \tag{2.7.8}$$

Thus, there exists a convergent subsequence of u_n, which is denoted u_n again:

$$\lim_{n\to\infty} u_n = u. \tag{2.7.9}$$

Since g is weakly lower semicontinuous, we obtain:

$$m = \lim_{n\to\infty} g(u_n) \leq g(u) \leq m. \tag{2.7.10}$$

Thus, the lemma is proved. □

Remark 2.7.1. The proof remains valid if we assume that the set $\{u : \varphi(u) \leq c\}$ is weakly precompact, i.e. contains a weakly convergent subsequence $u_n \rightharpoonup u$, rather than the strongly convergent one. However, the assumption about compactness of the embedding $i : H_1 \to H$ will be used later.

Theorem 2.7.1. *Assume that F is injective, the embedding $i : H_1 \to H$ is compact, $f = F(y)$, (2.7.2) holds, $||f_\delta - f|| \leq \delta$ and $g(u_\delta) = \inf_{u\in H} g(u)$. Then*

$$||u_\delta - y|| \leq \eta(\delta) \to 0 \quad as \quad \delta \to 0. \tag{2.7.11}$$

Proof. Using the equation $F(y) = f$, (2.7.7) and (2.7.8), we obtain:

$$||F(u_\delta) - F(y)|| \leq ||F(u_\delta) - f_\delta|| + ||f_\delta - f|| \leq \delta(2 + b\varphi(y)), \tag{2.7.12}$$

and

$$\varphi(u_\delta) \leq c. \tag{2.7.13}$$

Therefore (2.7.11) follows. Indeed, assuming that $||u_\delta - y|| \geq \varepsilon > 0$, choose a subsequence

$$u_n := u_{\delta_n} \to u \quad \text{as } \delta_n \to 0.$$

This is possible because i is compact and (2.7.13) holds. Then we get

$$\varepsilon \leq \lim_{n\to\infty} ||u_n - y|| = ||u - y||. \tag{2.7.14}$$

Let us prove that

$$F(u) = y. \tag{2.7.15}$$

It follows from (2.7.12) that

$$\lim_{n\to\infty} ||F(u_n) - f|| = 0.$$

Assumption (2.7.2) and weak lower semicontinuity of the norm in H imply

$$0 = \underline{\lim}_{n\to\infty} ||F(u_n) - f|| \geq ||F(u) - f||. \tag{2.7.16}$$

Thus (2.7.15) is verified.

Since F is injective, it follows from (2.7.15) that $u = y$. This contradicts to the inequality (2.7.14).

Theorem 2.7.1 is proved. □

Let us discuss now a *new discrepancy principle.*

Earlier the discrepancy principle for finding the regularization parameter $a = a(\delta)$ was proposed and justified for linear equations in the following form. One finds a minimizer $u_{a,\delta}$ to the functional

$$G(u) := ||F(u) - f_\delta||^2 + a||u||^2.$$

Then $a = a(\delta)$ is found as the solution (unique if F is a linear operator) of the equation

$$||F(u_{a,\delta}) - f_\delta|| = c\delta, \quad ||f_\delta|| > c\delta, \quad c = \text{const}, \quad c \in (1, 2). \tag{2.7.17}$$

The justification of the discrepancy principle consisted of the proof of the unique solvability of equation (2.7.17) for $a = a(\delta)$ and of the proof of the relation

$$\lim_{\delta\to 0} ||u_\delta - y|| = 0, \quad u_\delta := u_{a(\delta),\delta}. \tag{2.7.18}$$

Equation (2.7.17) is a nonlinear equation for a even in the case when $F(u) = Au$ is a linear operator.

To avoid solving this equation, we propose to take $a(\delta) = b\delta, \quad b = \text{const} > 0$.

We have proved that for any fixed $b = \text{const} > 0$ problem (2.7.5) has a solution u_δ and (2.7.11) holds, provided that F is injective, (2.7.2) holds and i is compact.

Therefore one may use the following relaxed version of the discrepancy principle. The usual discrepancy principle requires to solve the variational problem

$$||F(u) - f_\delta||^2 + a\varphi(u) = \min, \tag{2.7.19}$$

to find a minimizer $u_{a,\delta}$, then to find the regularization parameter $a = a(\delta)$ by solving the nonlinear equation for a:

$$||F(u_{a,\delta}) - f_\delta|| = c\delta, \quad c \in (1, 2), \quad c = \text{const}, \tag{2.7.20}$$

and then to prove that

$$u_\delta := u_{a(\delta),\delta}$$

satisfies (2.7.18).

The relaxed version of the discrepancy principle does not require solving nonlinear equation (2.7.20) for a, but allows one to take

$$a(\delta) = b\delta$$

with an arbitrary fixed $b = \text{const} > 0$.

One can also choose some b such that

$$\delta \leq ||F(u_{b\delta,\delta}) - f_\delta|| \leq c_1\delta, \qquad c_1 = \text{const} > 0, \tag{2.7.21}$$

where $c_1 > 1$ is an arbitrary fixed constant. It is easier numerically to find b such that (2.7.21) holds than to solve equation (2.7.20) for $a = a(\delta)$.

Chapter 3

DSM for well-posed problems

In this Chapter it is shown that every well-posed problem can be solved by a DSM which converges exponentially fast.

3.1 Every solvable well-posed problem can be solved by DSM

In this chapter we prove that every solvable equation

$$F(u) = 0, \tag{3.1.1}$$

where F satisfies assumptions (2.6.2) and (2.6.3), can be solved by a DSM (1.1.2) so that the three conditions (1.1.5) hold, and, in addition, the convergence of this DSM is exponentially fast:

$$||u(t) - u(\infty)|| \le re^{-c_1 t}, \quad ||F(u(t))|| \le ||F_0||e^{-c_1 t}, \tag{3.1.2}$$

where $c_1 = \text{const} > 0, \quad r = \text{const} > 0, \quad F_0 = F(u(0))$. We will specify c_1 and r later.

First, let us establish a general framework for a study of well-posed problems.

Assume that

$$(F'(u)\Phi(t,u), F(u)) \le -g_1(t)||F(u)||^2 \qquad \forall u \in H, \tag{3.1.3}$$

and

$$||\Phi(t,u)|| \le g_2(t)||F(u)||, \quad \forall u \in H, \tag{3.1.4}$$

where $g_1(t)$ and $g_2(t)$ are positive functions, defined on $\mathbb{R}_+ = [0,\infty)$, g_2 is continuous and $g_1 \in L^1(\mathbb{R}_+)$.

The assumption (3.1.3) can be generalized

$$||\Phi(t,u)|| \le g_2(t)||F(u)||^b, \quad b = \text{const} > 0. \tag{3.1.5}$$

Define the following function:

$$G(t) := g_2(t)e^{-\int_0^t g_1(s)ds}. \tag{3.1.6}$$

Assume:

$$\int_0^\infty g_1 ds = \infty, \quad G \in L^1(\mathbb{R}_+), \tag{3.1.7}$$

and

$$||F(u_0)|| \int_0^\infty G(t)dt \le R. \tag{3.1.8}$$

Let us formulate our first result.

Theorem 3.1.1. *Assume that (3.1.3), (3.1.4), (3.1.7) and (3.1.8) hold. Also assume that (2.6.2) holds for $j \le 1$ and (2.6.6) holds. Then problem (2.6.4) has a unique global solution $u(t)$, there exists $u(\infty)$, $F(u(\infty)) = 0$, and the following estimates hold:*

$$||u(t) - u(\infty)|| \le ||F(u_0)|| \int_t^\infty G(s)ds, \tag{3.1.9}$$

and

$$||F(u(t))|| \le ||F(u_0)||e^{-\int_0^t g_1(s)ds}. \tag{3.1.10}$$

Proof. From the assumption (2.6.6) it follows that there exists a unique local solution to problem (2.6.4). To prove that this solution is global we use Lemma 2.6.1 and establish estimate (2.6.8) for this solution. Let $g(t) := ||F(u(t))||$, and $\dot{g} = \frac{dg}{dt}$. Then, using (2.6.2) with $j = 1$, we get:

$$g\dot{g} = (F'(u)\dot{u}, F) = (F'(u)\Phi(t,u), F) \le -g_1(t)g^2. \tag{3.1.11}$$

Since $g \ge 0$, we obtain

$$g(t) \le g_0 e^{-\int_0^t g_1(s)ds}, \quad g_0 = ||F(u_0)||, \tag{3.1.12}$$

which is the inequality (3.1.10). From equation (2.6.4), inequality (3.1.4) and estimate (3.1.10) we derive the estimate:

$$||u(s) - u(t)|| \leq ||F(u_0)|| \int_t^s G(p)dp. \tag{3.1.13}$$

Since $G \in L^1(\mathbb{R}_+)$, it follows from (3.1.13) that estimate (2.6.8) holds, the limit

$$u(\infty) = \lim_{t \to \infty} u(t)$$

exists, and estimate (3.1.9) holds. Taking $t \to \infty$ in (3.1.10), using the first assumption (3.1.7) and the continuity of F, one obtains the relation $F(u(\infty)) = 0$. Theorem 3.1.1 is proved. □

Let us replace assumption (3.1.3) by a more general one:

$$(F'\Phi, F) \leq -g_1(t)||F(u)||^a, \quad a \in (0, 2). \tag{3.1.14}$$

Arguing as in the proof of Theorem 3.1.1 one gets the inequality:

$$g^{1-a}\dot{g} \leq -g_1(t),$$

so

$$0 \leq g(t) \leq \left[g^{2-a}(0) - (2-a) \int_0^t g_1(s)ds \right]^{\frac{1}{2-a}}. \tag{3.1.15}$$

If the first assumption (3.1.7) holds then (3.1.15) implies that $g(t) = 0$ for all $t \geq T$, where T is defined by the equation

$$\int_0^T g_1(s)ds = \frac{g^{2-a}(0)}{2-a}, \quad 0 < a < 2. \tag{3.1.16}$$

Thus

$$||F(u(t))|| = 0, \quad t \geq T. \tag{3.1.17}$$

Therefore $u(t)$ solves the equation $F(u) = 0$, and

$$||u(T) - u(0)|| \leq ||F(u_0)|| \int_0^T G(p)dp \leq ||G(u_0)|| \int_0^T g_2(s)ds. \tag{3.1.18}$$

From (3.1.18) and (3.1.8) it follows that $u(t) \in B(u_0, R)$ for all $t \geq 0$.

Let us formulate the results we have proved:

Theorem 3.1.2. *Assume (3.1.14), (3.1.8), (3.1.7), (2.6.6), (2.6.2) with* $j \leq 1$*, and let* T *be defined by equation (3.1.16). Then equation* $F(u) = 0$ *has a solution* $u \in B(u_0, R)$*, problem (2.6.4) has a unique global solution* $u(t) \in B(u_0, R)$*, and* $F(u(t)) = 0$ *for* $t \geq T$*.*

Assume now that the inequality (3.1.14) holds with $a > 2$ and the first assumption (3.1.7) holds. Then, arguing as in the proof of Theorem 3.1.1, one gets

$$0 \leq g(t) \leq \left[\frac{1}{g^{a-2}(0)} + (a-2)\int_0^t g_1(s)ds\right]^{-\frac{1}{a-2}} := h(t), \tag{3.1.19}$$

where $\lim_{t\to\infty} h(t) = 0$ because of (3.1.7).

Assume that

$$\int_0^\infty g_2(s)h(s)ds \leq R. \tag{3.1.20}$$

Then (2.6.4) and (3.1.4) imply

$$||u(t) - u(0)|| \leq R, \quad ||u(t) - u(\infty)|| \leq \int_t^\infty g_2(s)h(s)ds, \tag{3.1.21}$$

so

$$\lim_{t\to\infty} ||u(t) - u(\infty)|| = 0. \tag{3.1.22}$$

Thus, the following result is obtained.

Theorem 3.1.3. *Assume that (3.1.14) holds with* $a > 2$*, (3.1.4), (2.6.6), (3.1.7) and (3.1.20) hold. Then there exists a unique global solution* $u(t)$ *to (2.6.4),* $u(t) \in B(u_0, R)$*, there exists* $u(\infty)$ *and* $F(u(\infty)) = 0$*.*

In the remaining Sections of this Chapter we will use the above results in a number of problems of interest in applications.

In particular, we will use the following consequence of Theorem 3.1.1.

Assume that $g_1(t) = c_1 = \text{const} > 0$, $\quad g_2(t) = c_2 = \text{const} > 0$, so that (3.1.3) and (3.1.4) take the form

$$(F'(u)\Phi(t,u), F(u)) \leq -c_1||F(u)||^2, \quad \forall u \in H, \tag{3.1.23}$$

and

$$||\Phi(t,u)|| \leq c_2||F(u)||, \quad \forall u \in H. \tag{3.1.24}$$

Then conditions (3.1.7) are trivially satisfied, (3.1.8) takes the form

$$||F(u_0)||\frac{c_2}{c_1} := r \le R, \tag{3.1.25}$$

(3.1.13) implies

$$||u(\infty) - u(t)|| \le re^{-c_1 t}, \tag{3.1.26}$$

and

$$||u(t) - u(0)|| \le r, \tag{3.1.27}$$

and (3.1.12) yields:

$$||F(t)|| \le ||F(u_0)||e^{-c_1 t}. \tag{3.1.28}$$

Note that (3.1.2) follows from (3.1.26) and (3.1.28).

Let us formulate what we have just demonstrated.

Theorem 3.1.4. *Assume (3.1.23)-(3.1.25), (2.6.2) and (2.6.3). Then problem (2.6.4) has a unique global solution* $u(t)$*, this solution satisfies inequalities (3.1.26)-(3.1.28), so that there exists* $\lim_{t\to\infty} u(t) := u(\infty)$*, and* $F(u(\infty)) = 0$*. Thus, under the above assumptions equation* $F(u) = 0$ *has a solution* $u(\infty) = u(\infty; u_0)$*, possibly nonunique.*

By a *global* solution to (2.6.4) we mean the solution which exists for all $t \ge 0$.

Note that if $u(\infty; u_0)$ is taken as the initial data, then $u(\infty; u(\infty; u_0))$ does not have to be equal to $u(\infty; u_0)$.

Example 3.1.1. Let

$$\dot{u} = c(1+u^2)e^{-t}, \quad u(0) = u_0 = 1.$$

Then

$$\arctan u(t) - \arctan u(0) = c(1 - e^{-t}),$$
$$u(t) = \tan(\arctan u_0 + c - ce^{-t}).$$

Thus

$$u(\infty; u_0) = \tan(\arctan 1 + c) = \tan\left(\frac{\pi}{4} + c\right),$$

and

$$u(\infty; u(\infty; u_0)) = \tan(\arctan\tan\left(\frac{\pi}{4} + c\right) + c) = \tan\left(\frac{\pi}{4} + 2c\right) \ne u(\infty; u_0)$$

provided that $\frac{\pi}{4} + c \ne \frac{\pi}{4} + 2c + n\pi$.

In order that $u(\infty; u_0) = u(\infty; u(\infty; u_0))$ for the problem

$$\dot{u} = \Phi(t, u), \quad u(0) = u_0,$$

it is sufficient that $u(\infty; u_0)$ solves the equation $\Phi(\infty; u(\infty; u_0)) = 0$.

3.2 DSM and Newton-type methods

Consider the equation

$$F(u) = 0, \tag{3.2.1}$$

and the DSM for solving (3.2.1) of the form

$$\dot{u} = -[F'(u)]^{-1}F(u), \quad u(0) = u_0. \tag{3.2.2}$$

This is a particular case of (2.6.4) which makes sense if (2.6.3) holds, i.e. if the problem is well-posed in our terminology. Assumption (2.6.6) follows from (2.6.2) and (2.6.3) due to the differentiation formula:

$$([F'(u)]^{-1})' = -[F'(u)]^{-1}F''(u)[F'(u)]^{-1}, \tag{3.2.3}$$

which can be derived easily by differentiating the identity

$$[F'(u)]^{-1}F'(u) = I \tag{3.2.4}$$

with respect to u. We assume throughout the rest of this chapter that assumptions (2.6.2) - (2.6.3) hold. Let us apply Theorem 3.1.4 to problem (3.2.2). We have

$$(F'\Phi, F) = -||F||^2, \tag{3.2.5}$$

so that $c_1 = 1$ in condition (3.1.23). Furthermore, (3.1.24) holds with $c_2 = m(R)$, where $m(R)$ is the constant from (2.6.3). Finally, to ensure the existence of a solution to equation (3.2.1) let us assume that (3.1.25) holds:

$$||F(u_0)||\frac{c_2}{c_1} := r \le R, \quad \text{i.e.,} \quad ||F(u_0)||m(R) \le R. \tag{3.2.6}$$

Then Theorem 3.1.4 implies the following result.

Theorem 3.2.1. *Assume (2.6.2), (2.6.3) and (3.2.6). Then equation (3.2.1) has a solution in $B(u_0, R)$, problem (3.2.2) has a unique global solution $u(t) \in B(u_0, R)$, there exists the limit*

$$u(\infty) = u(\infty; u_0) \in B(u_0, R),$$

this limit solves the equation

$$F(u(\infty)) = 0,$$

and the DSM method (3.2.2) converges exponentially fast in the sense that estimates (3.1.26) and (3.1.28) hold.

Remark 3.2.1. Condition (3.2.6) is always satisfied for a suitable u_0 if one knows a priori that equation (3.2.1) has a solution y, $F(y) = 0$. Indeed, in this case one can choose u_0 sufficiently close to y and, by the continuity of F, condition (3.2.6) will be satisfied because $\lim_{u_0 \to y} ||F(u_0)|| = ||F(y)|| = 0$. Most of the classical theorems about convergence of the discrete Newton-type methods contain the assumption that u_0 is sufficiently close to a solution to equation (3.2.1).

Remark 3.2.2. If one formally discretizes equation (3.2.2), one gets an iterative scheme

$$u_{n+1} = u_n - h_n[F'(u_n)]^{-1}F(u_n), \quad u_0 = u_0, \tag{3.2.7}$$

$$t_{n+1} = t_n + h_n, \quad u_n = u(t_n). \tag{3.2.8}$$

If $h_n = 1$, the process, (3.2.7) reduces to the classical Newton's method.

$$u_{n+1} = u_n - [F'(u_n)]^{-1}F(u_n), \quad u_0 = u_0. \tag{3.2.9}$$

Various conditions sufficient for the convergence of this process are known (see [KA], [De]).

For a comparison with Theorem 3.2.1 let us formulate a theorem (see [De], p.157) about convergence of the classical Newton's method.

Theorem 3.2.2. *Assume*

$$||[F'(u_0)]^{-1}F(u_0)|| \le a, \quad ||[F'(u_0)]^{-1}|| \le b,$$

$$||F'(u) - F'(v)|| \le k||u - v||, \quad u, v \in B(u_0, R),$$

$$q := 2kab < 1 \text{ and } 2a < R.$$

Then F has a unique zero $y \in B(u_0, R)$, and

$$||u_n - y|| \le \frac{a}{2^{n-1}} q^{2^n - 1}, \quad 0 < q < 1. \tag{3.2.10}$$

A proof of Theorem 3.2.2 one can find in [De], p. 158. The proof is considerably more complicated than the proof of Theorem 3.2.1. The basic feature of the classical process (3.2.9) is its quadratic rate of convergence. This means that

$$||u_{n+1} - y|| \le c||u_n - y||^2$$

for all sufficiently large n. This rate is achieved for the process (3.2.7) with a constant step size $h_n = h$ only if $h = 1$. For the continuous analog (3.2.2) of the method (3.2.9) the rate of convergence is exponential, it is slower than quadratic.

3.3 DSM and the modified Newton's method

Consider equation (3.2.1) and the following DSM method for solving this equation:

$$\dot{u} = -[F'(u_0)]^{-1}F(u), \quad u(0) = u_0. \tag{3.3.1}$$

Let us check conditions of Theorem 3.1.4. Condition (3.1.24) is satisfied with $c_2 = m(R)$, where $m(R)$ is the constant from (2.6.3). Let us check condition (3.1.23). We have

$$\begin{aligned} &-(F'(u)[F'(u_0)]^{-1}F(u), F(u)) \\ &\quad = -([F'(u) - F'(u_0)][F'(u_0)]^{-1}F(u), F(u)) - \|F(u)\|^2, \end{aligned} \tag{3.3.2}$$

and

$$|[F'(u)-F'(u_0)][F'(u_0)]^{-1}F(u), F(u))| \leq M_2\|u-u_0\|m\|F(u)\|^2, \tag{3.3.3}$$

where M_2 is the constant from (2.6.2).

Let us assume that $\|u - u_0\| \leq R$, i.e. $u \in B(u_0, R)$, and choose R such that

$$mM_2R = \frac{1}{2}. \tag{3.3.4}$$

Then (3.3.2) - (3.3.4) imply that $c_1 = \frac{1}{2}$. Condition (3.1.25) takes the form

$$2\|F(u_0)\|m \leq \frac{1}{2mM_2},$$

i.e.

$$4m^2M_2\|F(u_0)\| \leq 1. \tag{3.3.5}$$

Theorem 3.1.4 yields the following result.

Theorem 3.3.1. *Assume (2.6.2),(2.6.3) and (3.3.5). Then equation (3.2.1) has a solution in* $B(u_0, R)$*, problem (3.3.1) has a unique global solution* $u(t) \in B(u_0, R)$*, there exists* $u(\infty) \in B(u_0, R)$, $F(u(\infty)) = 0$*, and the DSM method (3.3.1) converges to the solution* $u(\infty)$ *exponentially fast in the sense that estimates (3.1.26) and (3.1.28) hold.*

3.4 DSM and Gauss-Newton-type methods

Consider equation (3.2.1) and the following DSM method for solving this equation

$$\dot{u} = -T^{-1}A^*F(u), \quad u(0) = u_0, \tag{3.4.1}$$

where

$$A := F'(u), \quad T := A^*A, \tag{3.4.2}$$

A^* is the operator adjoint to A.

Let us apply Theorem 3.1.4. As always in this Chapter, we assume (2.6.2) and (2.6.3). Condition (3.1.23) takes the form:

$$-(F'T^{-1}A^*F, F) = -||F||^2, \tag{3.4.3}$$

because $A(A^*A)^{-1}A^* = I$, since $F'(u) := A$ is a boundedly invertible operator (see (2.6.3)). Thus, $c_1 = 1$.

Let us find c_2. We have

$$||T^{-1}A^*F|| \le ||T^{-1}||M_1||F||, \tag{3.4.4}$$

where $||A^*|| = ||A|| \le M_1$ by (2.6.2). Finally

$$||T^{-1}|| \le ||A^{-1}||||(A^*)^{-1}|| \le m^2, \tag{3.4.5}$$

where m is the constant from (2.6.3). Thus $c_2 = M_1 m^2$. Condition (3.1.25) takes the form

$$||F(u_0)||M_1 m^2 \le R. \tag{3.4.6}$$

Theorem 3.1.4 yields the following result.

Theorem 3.4.1. *Assume (2.6.2), (2.6.3) and (3.4.6). Then equation (3.2.1) has a solution in $B(u_0, R)$, problem (3.4.1) has a unique global solution $u(t)$, there exists $u(\infty)$, $F(u(\infty)) = 0$, and the DSM method (3.4.1) converges to $u(\infty)$ exponentially fast in the sense that estimates (3.1.26) and (3.1.28) hold.*

3.5 DSM and the gradient method

Again we want to solve equation (3.2.1). The DSM method we use is the following one:

$$\dot{u} = -A^*F(u), \quad u(0) = u_0. \tag{3.5.1}$$

As before, $A = F'(u)$, A is the adjoint to A. Condition (3.1.23) takes the form:

$$-(AA^*F, F) = -||A^*F||^2 \le -c_1||F||^2, \tag{3.5.2}$$

where $c_1 = m^2$. Here we have used the following estimates:

$$||A^*u|| \geq ||(A^*)^{-1}||^{-1}||u||, \quad ||(A^*)^{-1}|| = ||A^{-1}|| = m, \tag{3.5.3}$$

where m is the constant from (2.6.3). Condition (3.1.24) holds with $c_2 = M_1$, where M_1 is the constant from (2.6.2) and we have used the relation $||A^*|| = ||A||$. Condition (3.1.25) takes the form

$$||F(u_0)||m^2 M_1 \leq R. \tag{3.5.4}$$

Theorem 3.1.4 yields the following result.

Theorem 3.5.1. *Assume (2.6.2), (2.6.3) and (3.5.4). Then equation (3.2.1) has a solution in $B(u_0, R)$, problem (3.5.1) has a unique global solution $u(t) \in B(u_0, R)$, there exists $u(\infty)$, $F(u(\infty)) = 0$, and the DSM method (3.5.1) converges exponentially fast in the sense that estimates (3.1.26) and (3.1.28) hold.*

3.6 DSM and the simple iterations method

We want to solve equation (3.2.1) by the following DSM method:

$$\dot{u} = -F(u), \quad u(0) = u_0. \tag{3.6.1}$$

Let us assume that

$$F'(u) \geq c_1(R) > 0, \quad u \in B(u_0, R). \tag{3.6.2}$$

Condition (3.1.23) takes the form:

$$-(F'F, F) \leq -c_1||F||^2, \tag{3.6.3}$$

so $c_1 := c_1(R)$ is the constant from (3.6.2). Condition (3.1.24) holds with $c_2 = 1$. Condition (3.1.25) takes the form:

$$||F(u_0)||\frac{1}{c_1} \leq R. \tag{3.6.4}$$

Theorem 3.1.4 yields the following result.

Theorem 3.6.1. *Assume (2.6.2), (2.6.3) and (3.6.4). Then equation (3.2.1) has a solution in $B(u_0, R)$, problem (3.6.1) has a unique global solution $u(t) \in B(u_0, R)$, there exists $u(\infty)$, $F(u(\infty)) = 0$, and the DSM method (3.6.1) converges to $u(\infty)$ exponentially fast, i.e., inequalities (3.1.26) and (3.1.28) hold.*

3.7 DSM and minimization methods

Let $f : H \to \mathbb{R}_+$ be twice Fréchet differentiable function, $f \in C^2_{\text{loc}}$. We want to use DSM for global minimization of f. In many cases this gives a method for solving operator equation $F(u) = 0$. Indeed, if this equation has a solution y, $F(y) = 0$, then y is the global minimizer of the functional $f(u) = ||F(u)||^2$.

Let us consider the following DSM scheme:

$$\dot{u} = -\frac{h}{(f'(u), h)} f(u), \quad u(0) = u_0, \tag{3.7.1}$$

where $h \in H$ is some element for which $|(f'(u(t)), h)| > 0$ for $t \geq 0$, where $u(t)$ solves (3.7.1).

Condition (3.1.23) takes the form

$$-\frac{(f', h)}{(f', h)} f^2 = -f^2,$$

so $c_1 = 1$. Condition (3.1.24) is:

$$\frac{||h||}{|(f'(u), h)|} f(u) \leq c_2 f(u),$$

where c_2 depends on the choice of h and on $f(u)$. Choose

$$h = f'(u(t)).$$

Then (3.7.1) becomes

$$\dot{u} = -\frac{f'(u(t))}{||f'(u(t))||^2} f(u(t)), \quad u(0) = u_0. \tag{3.7.2}$$

Condition (3.1.24) is not satisfied, in general, and we assume that

$$\frac{f(u)}{||f'(u)||} \leq a f^b, \tag{3.7.3}$$

where a and b are positive constants.

Then (3.7.2) implies

$$\dot{f} = f'(u)\dot{u} = -f,$$

so

$$f(u(t)) = f_0 e^{-t}, \quad f_0 := f(u_0), \tag{3.7.4}$$

and

$$||\dot{u}|| \leq \frac{f}{||f'(u(t))||} \leq af_0^b e^{-bt} := c_3 e^{-bt}. \tag{3.7.5}$$

One has

$$||u||^{\cdot} \leq |\dot{u}|| \leq c_3 e^{-bt}, \tag{3.7.6}$$

so

$$||u(t) - u_0|| \leq \frac{c_3}{b}, \quad ||u(t) - u(\infty)|| \leq \frac{c_3}{b} e^{-bt}. \tag{3.7.7}$$

Condition (3.1.25) takes the form:

$$\frac{c_3}{b} \leq R, \quad \text{i.e.} \quad \frac{af_0^b}{b} \leq R. \tag{3.7.8}$$

We have proved the following result.

Theorem 3.7.1. *Assume that $f \in C^2_{loc}$, (3.7.3) and (3.7.8) hold. Then equation $f(u) = 0$ has a solution in $B(u_0, R)$, problem (3.7.2) has a unique global solution $u(t)$, $f(u(\infty)) = 0$, and estimates (3.7.7) hold.*

Remark 3.7.1. If $f(u) = ||F(u)||^2$ and H is the real Hilbert space, then $f'(u) = 2[F'(u)]^* F(u)$. Assume (2.6.3). Denote $A := F'(u)$. Then

$$||[A^*]^{-1}|| \leq m(R).$$

One has

$$||A^* F|| \geq ||[A^*]^{-1}||^{-1} ||F(u)|| = m^{-1} ||F(u)||.$$

Thus, if

$$\Phi = -\frac{f'}{||f'||^2} f, \quad f = ||F(u)||^2,$$

then

$$||\Phi|| \leq \frac{1}{||f'(u)||} f \leq \frac{m}{2||F||} f = \frac{m}{2} f^{\frac{1}{2}},$$

so that (3.7.2) holds with $b = \frac{1}{2}$. Therefore assumption (2.6.3) ensures convergence of the DSM (3.7.2) for global minimization of the functional $f(u) = ||F(u)||^2$ with an operator satisfying assumption (2.6.3), and the DSM converges exponentially fast.

3.8 Ulm's method

In [U] and [GW] the following method for solving equation $F(u) = 0$ is discussed under some assumptions on F, of which one is the existence of a bounded linear operator B in a Hilbert (or Banach) space such that

$$||I - BF'(u_0)|| < 1, \tag{3.8.1}$$

where $u_0 \in H$ is some element. The method consists of using the following iterative process

$$u_{n+1} = u_n - B_{n+1}F(u_n), \quad B_{n+1} = 2B_n - B_nF'(u_n)B_n. \tag{3.8.2}$$

As the initial approximation one takes u_0 and $B_0 = B$ from the assumption (3.8.1).

The aim of this short Section is to prove the convergence of a simplified continuous analog of the above method using Theorem 3.1.1. The continuous analog is of the form:

$$\dot{u} = -BF(u), \quad u(0) = u_0. \tag{3.8.3}$$

We have simplified the method by not considering the updates for B and assuming that

$$||I - BF'(u)|| \leq q < 1, \quad u \in B(u_0, R), \tag{3.8.4}$$

where $q \in (0, 1)$ is a number independent of u, and $R > 0$ is some number. Thus, B is an approximation to the inverse operator $[F'(u)]^{-1}$ in some sense. We wish to prove the convergence of the DSM (3.8.3). Our Φ from DSM (1.1.2) is

$$\Phi = -BF,$$

and we wish to verify conditions of Theorem 3.1.1.

The DSM (3.8.3) is similar to the Newton method. The main difference between the two methods is in taking an approximate inverse B in the sense (3.8.4) in place of the exact inverse $[F'(u)]^{-1}$ in the Newton method.

Let us verify the conditions of Theorem 3.1.1. Condition (3.1.3) is easy to verify:

$$-(F'BF, F) \leq -(1 - q)||F||^2,$$

where we have used the assumption (3.8.4). Thus,

$$g_1(t) = c_1 = 1 - q > 0,$$

and condition (3.1.3) holds.

Condition (3.1.4) obviously holds with

$$g_2(t) = ||B|| := c_2.$$

Condition (3.1.8) takes the form (3.1.25), that is:

$$||F(u_0)||\frac{c_2}{c_1} \leq R. \tag{3.8.5}$$

If condition (3.8.5) holds, then,by Theorem 3.1.1, the DSM (3.8.3) is justified, i.e., the conclusions (1.1.5) are valid.

If equation (3.8.1) has a solution, then condition (3.8.5) is always satisfied if the initial approximation u_0 is taken sufficiently close to the solution.

Chapter 4

DSM and linear ill-posed problems

In this chapter we prove that any solvable linear operator equation can be solved by a DSM and by a convergent iterative process.

4.1 Equations with bounded operators

Consider the equation

$$Au = f, \tag{4.1.1}$$

where A is a bounded linear operator in a Hilbert space H. Assume that there exists a solution to (4.1.1), that $\mathcal{N} = \mathcal{N}(A)$ is the null-space of A, and denote by y the unique minimal-norm solution, i.e. the solution $y \perp \mathcal{N}$. Let $T := A^*A$, $\quad T_a := T + aI$, $\quad I$ is the identity operator. We assume that problem (4.1.1) is ill-posed in the sense that the range $\mathcal{R}(A)$ of the operator A is not closed.

If $\mathcal{R}(A)$ is closed, i.e. $\mathcal{R}(A) = \overline{\mathcal{R}(A)}$, but A is not injective, then one may define an operator A_1, from $H_1 := H \ominus \mathcal{N}$ onto $\mathcal{R}(A)$. This operator will be continuous, injective and, by the closed graph theorem , the inverse operator A_1^{-1} is continuous from $\mathcal{R}(A)$ into H_1. Thus, the ill-posedness which is due to the lack of injectivity is not a problem if $\mathcal{R}(A)$ is closed. Similarly, if A is not surjective but $\mathcal{R}(A)$ is closed then small perturbations f_δ of f which do not throw f_δ out of $\mathcal{R}(A)$, i.e. $f_\delta \in \mathcal{R}(A) = \overline{\mathcal{R}(A)}$, $||f_\delta - f|| \leq \delta$, lead to a small perturbation of the minimal-norm solution because A_1^{-1} is continuous. However, if $\mathcal{R}(A) \neq \overline{\mathcal{R}(A)}$, then the inverse operator A_1^{-1} from $\mathcal{R}(A)$ into H_1 is unbounded. Indeed, if A_1^{-1} is bounded

then

$$||u|| = ||A^{-1}Au|| \le c||Au||, \quad c = ||A^{-1}|| = \text{const} < \infty.$$

If

$$||Au_n - Au_m|| \to 0, \quad m, n \to \infty,$$

then the above inequality implies $||u_n - u_m|| \to 0, \quad m, n \to \infty$, so $u_n \to u$, and, by continuity of A, $Au_n \to Au$. Therefore $\mathcal{R}(A)$ is closed, contrary to our assumption.

The case when $\mathcal{R}(A)$ is not closed leads to difficulties in solving equation (4.1.1) because small perturbations of f may lead to large perturbations of u or may lead to an equation which has no solutions.

Let us assume that $\mathcal{R}(A) \neq \overline{\mathcal{R}(A)}$, f_δ is given, $||f_\delta - f|| \le \delta$, and

$$Ay = f, \quad y \perp \mathcal{N}.$$

How does one calculate by a DSM a stable approximation u_δ to y? That is, how does one calculate u_δ such that:

$$\lim_{\delta \to 0} ||u_\delta - y|| = 0. \tag{4.1.2}$$

There are several versions of DSM for solving ill-posed equation (4.1.1). One of these versions is

$$\dot{u} = -u + T_{a(t)}^{-1} A^* f, \quad u(0) = u_0, \tag{4.1.3}$$

where

$$a(t) > 0; \quad a(t) \searrow 0 \text{ as } t \to \infty, \tag{4.1.4}$$

and $\searrow$ denotes monotone decay to zero.

Theorem 4.1.1. *Problem (4.1.3) has a unique global solution $u(t)$, there exists $u(\infty)$, $\quad A(u(\infty)) = f, \quad u(\infty) = y \perp \mathcal{N}$.*

The case of noisy data f_δ will be discussed after the proof of Theorem 4.1.1.

Proof of Theorem 4.1.1. The unique solution of (4.1.3) is:

$$u(t) = u_0 e^{-t} + \int_0^t e^{-(t-s)} T_{a(s)}^{-1} A^* Ay \, ds.$$

Clearly, $\lim_{t \to \infty} ||u_0 e^{-t}|| = 0$. We claim that

$$\lim_{t \to \infty} \int_0^t e^{-(t-s)} T_{a(s)}^{-1} Ty \, ds = y. \tag{4.1.5}$$

This claim follows from two lemmas.

Lemma 4.1.1. *If g is a continuous function on $[0,\infty)$ with values in H and $g(\infty)$ exists, then*

$$\lim_{t\to\infty}\int_0^t e^{-(t-s)}g(s)\,ds = g(\infty). \tag{4.1.6}$$

Lemma 4.1.2. *One has*

$$\lim_{a\to 0} T_a^{-1}Ty = y, \quad a>0, \quad y\perp\mathcal{N}. \tag{4.1.7}$$

Proof of Lemma 4.1.1. For any fixed τ, however large, one has

$$\lim_{t\to\infty}\int_0^\tau e^{-(t-s)}g(s)\,ds = 0.$$

If τ is sufficiently large then $|g(s)-g(\infty)|<\eta, \quad \forall s\geq\tau$, where $\eta>0$ is an arbitrary small number. The conclusion (4.1.6) follows now from the relation $\lim_{t\to\infty}\int_\tau^t e^{-(t-s)}g(\infty)\,ds = g(\infty)$.

Lemma 4.1.1 is proved. □

Proof of Lemma 4.1.2. Using the spectral theorem for the selfadjoint operator T and denoting by E_s the resolution of the identity, corresponding to T, one gets:

$$\lim_{a\to 0}||T_a^{-1}Ty-y||^2 = \lim_{a\to 0}\int_0^{||T||}\frac{a^2 d(E_s y,y)}{(a+s)^2} = ||P_{\mathcal{N}}y||^2 = 0, \tag{4.1.8}$$

where

$$P_{\mathcal{N}} = E_0 - E_{-0}$$

is the orthoprojector onto the null space $\mathcal{N}$ of operator T. Because $\mathcal{N}(T)=\mathcal{N}(A)$ we use the same letter $\mathcal{N}$ as for the null-space of A.

Lemma 4.1.2 is proved. □

Since $\lim_{t\to\infty}a(t)=0$, Theorem 4.1.1 is proved. □

Consider now the case of noisy data f_δ. Then we solve the problem

$$\dot{v} = -v + T_{a(t)}^{-1}A^*f_\delta, \quad v(0)=u_0, \tag{4.1.9}$$

stop integration at a stopping time t_δ, denote $v(t_\delta):=u_\delta$, and prove (4.1.2) for a suitable choice of t_δ. Let $u(t)-v(t):=w(t)$. Then

$$\dot{w} = -w + T_{a(t)}^{-1}A^*(f-f_\delta), \quad w(0)=0. \tag{4.1.10}$$

By (2.2.12) we have

$$||T_a^{-1}A^*|| \le \frac{1}{2\sqrt{a}}. \tag{4.1.11}$$

Since

$$w(t) = \int_0^t e^{-(t-s)} T_{a(s)}^{-1} A^*(f - f_\delta) ds,$$

one has

$$||w(t)|| \le \int_0^t e^{-(t-s)} \frac{\delta}{2\sqrt{a(s)}} ds \le \frac{\delta}{2\sqrt{a(t)}}. \tag{4.1.12}$$

Therefore, if t_δ is chosen so that

$$\lim_{\delta \to 0} \frac{\delta}{2\sqrt{a(t_\delta)}} = 0, \quad \lim_{\delta \to 0} t_\delta = \infty, \tag{4.1.13}$$

then (4.1.12) and Theorem 4.1.1 imply

Theorem 4.1.2. *Assume that (4.1.13) holds and equation (4.1.1) is solvable. Then $u_\delta := v(t_\delta)$, where $v(t)$ is the unique solution to (4.1.9), satisfies (4.1.2).*

Remark 4.1.1. There are many a priori choices of the stopping times t_δ satisfying (4.1.13). No optimization of the choice of t_δ has been made. It is an open problem to propose such an optimization.

Let us propose a posteriori choice of t_δ based on a discrepancy-type principle. We choose t_δ from the equation

$$||AT_{a(t)}^{-1}A^* f_\delta - f_\delta|| = c\delta, \quad c = \text{const} \in (1, 2). \tag{4.1.14}$$

Denote $AA^* := Q$. We have

$$a^2(t) \int_0^{||Q||} \frac{1}{[s + a(t)]^2} d(F_s f_\delta, f_\delta) = c^2 \delta^2, \tag{4.1.15}$$

where F_s is the resolution of the identity corresponding to the selfadjoint operator Q. Equation (4.1.15) for a has a unique solution a_δ and t_δ is found uniquely from the equation

$$a_\delta = a(t). \tag{4.1.16}$$

This equation has a unique solution $t = t_\delta$ because $a(t)$ is monotone. Equation (4.1.15) considered as an equation for a has a unique solution $a = a_\delta$. This was proved in Theorem 2.2.5. Since $\lim_{\delta \to 0} a_\delta = 0$, it follows that $\lim_{\delta \to 0} t_\delta = \infty$.

This allows us to prove the following result.

Theorem 4.1.3. *Equation (4.1.14) has a unique solution* $t = t_\delta$, *and* $\lim_{\delta\to 0} t_\delta = \infty$. *The element* $u_\delta := v(t_\delta)$, *where* v *solves (4.1.9) and* t_δ *is the solution of (4.1.14), satisfies (4.1.2), provided that*

$$\lim_{t\to\infty} \frac{\dot{a}(t)}{a^2(t)} = 0. \tag{4.1.17}$$

Proof of Theorem 4.1.3. By Theorem 2.2.5 we have

$$\lim_{\delta\to 0} ||T_{a_\delta}^{-1} A^* f_\delta - y|| = 0. \tag{4.1.18}$$

Denote

$$w_\delta(t_\delta) := T_{a_\delta}^{-1} A^* f_\delta.$$

We have

$$||v(t_\delta) - y|| \le ||v(t_\delta) - w_\delta(t_\delta)|| + ||w_\delta(t_\delta) - y||. \tag{4.1.19}$$

By (4.1.18),

$$\lim_{\delta\to 0} ||w_\delta(t_\delta) - y|| = 0. \tag{4.1.20}$$

Let us check that

$$\lim_{\delta\to 0} ||v(t_\delta) - w_\delta(t_\delta)|| = 0. \tag{4.1.21}$$

We have

$$v(t_\delta) = u_0 e^{-t_\delta} + \int_0^{t_\delta} e^{-(t_\delta - s)} w_\delta(s)ds, \quad w_\delta(s) := T_{a(s)}^{-1} A^* f_\delta. \tag{4.1.22}$$

Since

$$\lim_{\delta\to 0} t_\delta = \infty,$$

we have

$$\lim_{\delta\to 0} ||u_0|| e^{-t_\delta} = 0. \tag{4.1.23}$$

Furthermore,

$$\begin{aligned} \int_0^t e^{-(t-s)} w_\delta(s)ds &= e^{-(t-s)} w_\delta(s)\Big|_0^t - \int_0^t e^{-(t-s)} \dot{w}_\delta(s)ds \\ &= w_\delta(t) - e^{-t} w_\delta(0) - \int_0^t e^{-(t-s)} \dot{w}_\delta(s)ds. \end{aligned} \tag{4.1.24}$$

Therefore,

$$\begin{aligned}\lim_{\delta\to 0}||v(t_\delta)-w_\delta(t_\delta)|| &= \lim_{\delta\to 0}||\int_0^{t_\delta} e^{-(t_\delta-s)}\dot{w}_\delta(s)ds|| \\ &\le \lim_{\delta\to 0}\int_0^{t_\delta} e^{-(t_\delta-s)}||\dot{w}_\delta(s)||ds. \end{aligned} \tag{4.1.25}$$

If

$$\lim_{t\to\infty}||\dot{w}_\delta(t)||=0, \tag{4.1.26}$$

then, by Lemma 4.1.1, the desired relation (4.1.20) follows from (4.1.25). Let us verify (4.1.26). By the spectral theorem we have

$$w_\delta(t)=\int_0^b \frac{dE_\lambda A^* f_\delta}{a(t)+\lambda}, \quad b=||T||, \tag{4.1.27}$$

where E_λ is the resolution of the identity corresponding to the selfadjoint operator T. Thus

$$\dot{w}_\delta(t)=-\dot{a}(t)\int_0^b \frac{dE_\lambda A^* f_\delta}{[a(t)+\lambda]^2}=-\frac{\dot{a}(t)}{a^2(t)}\int_0^b \frac{a^2(t)dE_\lambda A^* f_\delta}{[a(t)+\lambda]^2}. \tag{4.1.28}$$

Consequently,

$$\lim_{t\to\infty}||\dot{w}_\delta(t)||\le \lim_{t\to\infty}\frac{|\dot{a}(t)|}{a^2(t)}||A^* f_\delta||^2=0, \tag{4.1.29}$$

where the assumption (4.1.17) was used.

Theorem 4.1.3 is proved. □

Remark 4.1.2. In the proof of Theorem 4.1.3 we have assumed that A is bounded, so that $A^* f_\delta$ makes sense for any $f_\delta \in H$. If A is unbounded, then our argument can be modified so that the conclusion of Theorem 4.1.3 still be true. Namely, we have

$$T_a^{-1}A^* f_\delta = A^* Q_a^{-1} f_\delta = UQ^{\frac{1}{2}}Q_a^{-1} f_\delta, \quad Q=AA^*=Q^*, \tag{4.1.30}$$

where U is an isometry and $a>0$.

Let F_λ be the resolution of the identity corresponding to Q. Then

$$\dot{w}_\delta(t)=U\int_0^\infty \frac{-\dot{a}(t)\lambda^{\frac{1}{2}}dF_\lambda f_\delta}{[a(t)+\lambda]^2}, \tag{4.1.31}$$

so

$$\begin{aligned} ||\dot{w}_\delta(t)||^2 &\le \int_0^\infty \frac{|\dot{a}(t)|^2 \lambda d(F_\lambda f_\delta, f_\delta)}{[a(t)+\lambda]^4} \\ &\le \frac{|\dot{a}(t)|}{a^3(t)} \int_0^\infty \frac{\lambda a^3(t) d(F_\lambda f_\delta, f_\delta)}{[a(t)+\lambda]^4} \le \frac{|\dot{a}(t)|}{a^3(t)} ||f_\delta||^2. \qquad (4.1.32) \end{aligned}$$

Assume that

$$\lim_{t\to\infty} \frac{|\dot{a}(t)|}{a^3(t)} = 0. \qquad (4.1.33)$$

Then (4.1.32) implies (4.1.26) and, therefore, the conclusion of Theorem 4.1.3 remains valid.

Remark 4.1.3. Condition (4.1.17) holds, for instance, if

$$a(t) = \frac{c_0}{(c_1+t)^b}, \quad c_1, c_0 > 0, \quad b \in (0,1), \qquad (4.1.34)$$

and (4.1.33) holds if $b \in (0, \frac{1}{2})$.

Let us formulate another version of the discrepancy principle. Assume that

$$\lim_{t\to\infty} \left[e^t a(t) \left|\left| Q^{-1}_{a(t)} f_\delta \right|\right| \right] = \infty, \quad Q_a := AA^* + aI. \qquad (4.1.35)$$

Theorem 4.1.4. *Assume (4.1.35), (4.1.4) and*

$$\lim_{t\to\infty} \frac{|\dot{a}(t)|}{a(t)} = 0. \qquad (4.1.36)$$

Then equation

$$\int_0^t e^{-(t-s)} a(s) \left|\left| Q^{-1}_{a(s)} f_\delta \right|\right| ds = c\delta, \quad c \in (1,2), \quad ||f_\delta|| > c\delta, \quad (4.1.37)$$

has a unique solution t_δ and $v_\delta := v_\delta(t_\delta)$ converges to y, where $v_\delta(t)$ is the solution to (4.1.9).

Remark 4.1.4. If $f_\delta \notin \mathcal{R}(A)$, then assumption (4.1.35) is satisfied. Indeed

$$\lim_{a\to 0} \left|\left| a Q_a^{-1} f_\delta \right|\right|^2 = \lim_{a\to 0} \int_0^\infty \frac{a^2 d(F_\lambda f_\delta, f_\delta)}{(a+\lambda)^2} = ||Pf_\delta||^2 > 0, \qquad (4.1.38)$$

where P is the orthoprojector onto the null space of the operator Q, $\mathcal{N}(Q) = \mathcal{N}(A^*)$, and

$$||Pf_\delta|| > 0 \quad \text{if } f_\delta \notin \mathcal{R}(A). \qquad (4.1.39)$$

Indeed $f \in \mathcal{R}(A)$ and $\overline{\mathcal{R}(A)} = \mathcal{N}(A)^\perp$.

Proof of Theorem 4.1.4. Let

$$h(t) := a(t) \left\| Q_{a(t)}^{-1} f_\delta \right\| := a(t)g(t). \tag{4.1.40}$$

If (4.1.35) holds, then, by the L'Hospital rule, we have

$$\lim_{t\to\infty} \frac{e^t h(t)}{\int_0^t e^s h(s)ds} = 1, \tag{4.1.41}$$

provided that

$$\lim_{t\to\infty} \frac{\dot{h}(t)}{h(t)} = 0. \tag{4.1.42}$$

Relation (4.1.42) holds if

$$\lim_{t\to\infty} \frac{(h^2)^{\cdot}}{h^2(t)} = 0. \tag{4.1.43}$$

Relation (4.1.43) holds if (4.1.36) holds. Indeed,

$$\frac{(h^2)^{\cdot}}{h^2(t)} = \frac{(a^2)^{\cdot}}{a^2} + \frac{(g^2)^{\cdot}}{g^2}. \tag{4.1.44}$$

If (4.1.36) holds then

$$\lim_{t\to\infty} \frac{(a^2)^{\cdot}}{a^2} = 0. \tag{4.1.45}$$

Moreover, (4.1.36) implies

$$\lim_{t\to\infty} \frac{(g^2)^{\cdot}}{g^2(t)} = 0. \tag{4.1.46}$$

Indeed,

$$g^2(t) = \int_0^\infty \frac{d\rho(\lambda)}{[a(t)+\lambda]^2}, \quad d\rho(\lambda) := d(F_\lambda f_\delta, f_\delta), \tag{4.1.47}$$

so

$$(g^2)^{\cdot} = -2\dot{a} \int_0^\infty \frac{d\rho}{[a(t)+\lambda]^3} = 2\frac{|\dot{a}|}{a} \int_0^\infty \frac{a(t)d\rho}{[a(t)+\lambda]^3} \le 2\frac{|\dot{a}|}{a} g^2(t). \tag{4.1.48}$$

Thus, (4.1.46) follows from (4.1.36).

Relation (4.1.41) is equivalent to

$$\lim_{t\to\infty} \frac{\int_0^t e^{-(t-s)} h(s)ds}{h(t)} = 1, \tag{4.1.49}$$

which holds if (4.1.35), (4.1.36) and (4.1.4) hold. It follows from (4.1.49) that equation (4.1.37) can be written as

$$a(t)||Q_{a(t)}^{-1} f_\delta|| = c_1 \delta, \tag{4.1.50}$$

where

$$c_1 = c[1 + o(1)], \quad t \to \infty. \tag{4.1.51}$$

Thus

$$c_1 \in (1, 2) \quad \text{as} \qquad t \to \infty. \tag{4.1.52}$$

Consider the equation

$$||aQ_a^{-1} f_\delta||^2 = ||f_\delta - QQ_a^{-1} f_\delta||^2 = ||f_\delta - AT_a^{-1} A^* f_\delta||^2 = c_1^2 \delta^2. \tag{4.1.53}$$

This equation has a unique solution

$$a = a_\delta, \quad \lim_{\delta\to 0} a_\delta = 0, \tag{4.1.54}$$

as was proved in Theorem 2.2.1, where it was also proved that relation (2.2.7) holds with $u_\delta := T_{a_\delta}^{-1} A^* f_\delta$. Given a_δ, one finds a unique t_δ from the equation

$$a(t) = a_\delta. \tag{4.1.55}$$

This equation is uniquely solvable for all sufficiently small $\delta > 0$ because $a(t)$ is monotone and (4.1.54) holds. The solution t_δ of (4.1.55) has the property

$$\lim_{\delta\to 0} t_\delta = \infty. \tag{4.1.56}$$

By (4.1.49) equation (4.1.37) is asymptotically, as $\delta \to 0$, equivalent to equation (4.1.55). Therefore, its solution t_δ satisfies (4.1.56), and relation (2.2.7) holds with $u_\delta := v_\delta(t_\delta)$, because

$$\lim_{\delta\to 0} v_\delta(t_\delta) = \lim_{\delta\to 0} T_{a_\delta}^{-1} A^* f_\delta, \tag{4.1.57}$$

due to (4.1.49) and (4.1.56).

Theorem 4.1.4 is proved. □

4.2 Another approach

In this Section we develop another approach to solving ill-posed equation (4.1.1). The operator A in Theorems 4.1.1 - 4.1.3 may be unbounded.

Let us first explain the main idea: under suitable assumptions the inverse of an operator can be calculated by solving a Cauchy problem. Let B be a bounded linear operator which has a bounded inverse, and assume that

$$\lim_{t\to\infty} ||e^{Bt}|| = 0. \tag{4.2.1}$$

For example, a sufficient condition for (4.2.1) to hold is $B = B_1 + iB_2$, B_1 and B_2 are selfadjoint operators and $B_1 \le -cI$, $c = \text{const} > 0$. The operator e^{Bt} is well defined not only for bounded operators B but for generators of C_0-semigroups (see [P]). One has

$$\int_0^t e^{Bs} ds = B^{-1}(e^{Bt} - I), \tag{4.2.2}$$

and if (4.2.1) holds then

$$-\lim_{t\to\infty} \int_0^t e^{Bs} ds = B^{-1}. \tag{4.2.3}$$

The Cauchy problem:

$$\dot{W} = BW + I, \quad W(0) = 0, \tag{4.2.4}$$

has a unique solution

$$W(t) = \int_0^t e^{Bs} ds. \tag{4.2.5}$$

Therefore the problem

$$\dot{u} = Bu + f, \quad u(0) = 0, \tag{4.2.6}$$

has a unique solution

$$u = W(t)f = \int_0^t e^{Bs} dsf = \int_0^t e^{B(t-s)} dsf, \tag{4.2.7}$$

so

$$-\lim_{t\to\infty} u(t) = B^{-1}f. \tag{4.2.8}$$

Therefore, if A in (4.1.1) is boundedly invertible operator, such that

$$\lim_{t\to\infty} ||e^{At}|| = 0,$$

then one can solve equation (4.1.1) by solving the Cauchy problem

$$\dot{u} = Au - f, \quad u(0) = 0, \tag{4.2.9}$$

which has a unique global solution

$$u(t) = -\int_0^t e^{A(t-s)}ds\ f, \tag{4.2.10}$$

and then calculating the solution $y = A^{-1}f$ by the formula:

$$\lim_{t\to\infty} u(t) = A^{-1}f = y. \tag{4.2.11}$$

If A is not boundedly invertible, the above idea is also useful as we are going to show.

Let us assume that A is a linear selfadjoint densely defined operator, $\mathcal{N}$ is its null-space, E_s its resolution of the identity, and consider the problem

$$\dot{u}_a = iA_{ia}u_a - if, \quad u(0) = 0, \quad a = \text{const} > 0, \quad A_{ia} := A + iaI. \tag{4.2.12}$$

For any $a > 0$ the inverse A_{ia}^{-1} is a bounded operator,

$$||A_{ia}^{-1}|| \leq \frac{1}{a}.$$

Moreover,

$$\lim_{t\to\infty} ||e^{iA_{ia}t}|| = \lim_{t\to\infty} e^{-at} = 0, \tag{4.2.13}$$

so that (4.2.1) holds with $B = iA_{ia}$. Problem (4.2.12) has a unique global solution and, by (4.2.8), we have

$$\lim_{t\to\infty} u_a(t) = i(iA_{ia})^{-1}f = i(iA_{ia})^{-1}Ay, \quad y \perp \mathcal{N}. \tag{4.2.14}$$

Moreover,

$$\lim_{a\to 0} ||i(iA_{ia})^{-1}Ay - y|| = 0. \tag{4.2.15}$$

Indeed,

$$\lim_{a\to 0} \eta^2(a) := \lim_{a\to 0} ||A_{ia}^{-1}Ay - y||^2 = \lim_{a\to 0} \int_{-\infty}^{\infty} \frac{a^2 d(E_s y, y)}{s^2 + a^2} = ||P_{\mathcal{N}}y||^2 = 0. \tag{4.2.16}$$

We have proved the following Theorem.

Theorem 4.2.1. *Assume that A is a linear, densely defined selfadjoint operator, equation $Au = f$ is solvable (possibly nonuniquely) and y is its minimal-norm solution, $y \perp \mathcal{N} = \mathcal{N}(A)$. Then*

$$y = \lim_{a\to 0} \lim_{t\to\infty} u_a(t), \tag{4.2.17}$$

where $u_a(t) = u_a(t; f)$ is the unique solution to (4.2.12).

Remark 4.2.1. Practically one integrates (4.2.12) on a finite interval $[0, \tau]$ and, for a chosen $\tau > 0$ one takes $a = a(\tau) > 0$ such that

$$\lim_{\tau\to\infty} u_{a(\tau)}(\tau) = y. \tag{4.2.18}$$

Relation (4.2.18) holds if the following conditions are valid:

$$\lim_{\tau\to\infty} a(\tau) = 0, \quad \lim_{\tau\to\infty} \frac{e^{-a(\tau)\tau}}{a(\tau)} = 0. \tag{4.2.19}$$

For example, one may take $a(\tau) = \tau^{-\gamma}, \quad \gamma \in (0, 1)$. To check that (4.2.19) implies (4.2.18), we note that

$$||u_a(t) - u_a(\infty)|| \leq \frac{e^{-at}}{a}, \tag{4.2.20}$$

as follows from the proof of Theorem 4.2.1. Therefore

$$||u_{a(\tau)}(\tau)-y|| \leq ||u_{a(\tau)}(\tau)-u_{a(\tau)}(\infty)||+||u_{a(\tau)}(\infty)-y|| \to 0 \text{ as } \tau \to \infty, \tag{4.2.21}$$

provided that conditions (4.2.19) hold.

Remark 4.2.2. We set $u(0) = u_0$ in (4.2.12) but similarly one can treat the case $u(0) = u_0$ with an arbitrary u_0.

Let us consider the case of noisy data f_δ, $||f_\delta - f|| \leq \delta$. We have

$$\begin{aligned} ||u_a(t; f) - u_a(t; f_\delta)|| &\leq ||\int_0^t e^{i(A+ia)(t-s)} ds(-i)(f - f_\delta)|| \\ &\leq \frac{\delta}{a}||e^{i(A+ia)t} - I|| \leq \frac{2\delta}{a}, \end{aligned} \tag{4.2.22}$$

where $u_a(t, f_\delta)$ solves (4.2.12) with f_δ replacing f. Thus:

$$||u_a(t; f_\delta) - y|| \leq \frac{\delta}{a} + ||u_a(t; f) - y||. \tag{4.2.23}$$

We estimate:

$$||u_a(t;f) - y|| \le \frac{e^{at}}{a}||f|| + \eta(a), \quad \lim_{a\to 0} \eta(a) = 0. \tag{4.2.24}$$

From (4.2.23) - (4.2.24) one concludes that

$$\lim_{\delta\to 0} ||u_{a(\delta)}(t_\delta; f_\delta) - y|| = 0, \tag{4.2.25}$$

provided that

$$\lim_{\delta\to 0} a(\delta) = 0, \quad \lim_{\delta\to 0} t_\delta = \infty, \quad \lim_{\delta\to 0} \frac{\delta}{a(\delta)} = 0, \quad \lim_{\delta\to 0} \frac{e^{-t_\delta a(\delta)}}{a(\delta)} = 0. \tag{4.2.26}$$

Let us formulate the result.

Theorem 4.2.2. *Assume that equation $Au = f$ is solvable, $Ay = f$, $y \perp \mathcal{N}$, A is a linear selfadjoint densely defined operator in a Hilbert space H, $||f_\delta - f|| \le \delta$, and $u_a(t; f_\delta)$ solves (4.2.12) with f_δ replacing f. If (4.2.26) holds, then (4.2.25) holds.*

So far we have assumed that a = const. Let us take $a = a(t)$ and assume

$$0 < a(t) \searrow 0, \quad \int_0^\infty a(s)ds = \infty, \quad a' + a^2 \in L^1[0,\infty). \tag{4.2.27}$$

Consider the following DSM problem:

$$\dot{u} = i[A + ia(t)]u - if, \quad u(0) = 0. \tag{4.2.28}$$

Problem (4.2.28) has a unique solution

$$u(t) = \int_0^t e^{iA(t-s) - \int_s^t a(p)dp} ds(-if). \tag{4.2.29}$$

For exact data f we prove the following result.

Theorem 4.2.3. *If (4.2.27) holds, then*

$$\lim_{t\to\infty} ||u(t) - y|| = 0, \tag{4.2.30}$$

where $u(t)$ solves (4.2.28).

Proof of Theorem 4.2.3. Substitute $f = Ay$ in (4.2.29) and integrate by parts to get

$$u(t) = e^{iA(t-s)-\int_s^t adp} y\Big|_0^t - \int_0^t e^{iA(t-s)} a(s) e^{-\int_s^t adp} ds\ y.$$

Thus

$$u(t) = y - e^{iAt-\int_0^t adp} y - \int_0^t e^{iA(t-s)} a(s) e^{-\int_s^t adp} ds\ y, \tag{4.2.31}$$

and

$$||u(t) - y|| \le e^{-\int_0^t adp} ||y|| + || \int_0^t e^{iA(t-s)} a(s) e^{-\int_s^t adp} ds\ y|| := J_1 + J_2. \tag{4.2.32}$$

Assumption (4.2.27) implies $\lim_{t\to\infty} J_1 = 0$. Let us prove that

$$\lim_{t\to\infty} J_2 = 0.$$

Using the spectral theorem and denoting by E_λ the resolution of the identity corresponding to A, we get

$$J_2^2 = \int_\infty^\infty d(E_\lambda y, y) \left| \int_0^t e^{i\lambda(t-s)} a(s) e^{-\int_s^t adp} ds \right|^2. \tag{4.2.33}$$

We *claim* that assumptions (4.2.27) imply:

$$\lim_{t\to\infty} \int_0^t e^{i\lambda(t-s)} a(s) e^{-\int_s^t adp} ds = 0 \quad \forall \lambda \neq 0, \tag{4.2.34}$$

while for $\lambda = 0$ we have

$$\lim_{t\to\infty} \int_0^t a(s) e^{-\int_s^t adp} ds = \lim_{t\to\infty} (1 - e^{-\int_0^t adp}) = 1. \tag{4.2.35}$$

To verify (4.2.34) denote the integral in (4.2.34) by J_3, integrate by parts and get:

$$J_3 = \frac{e^{i\lambda(t-s)}}{-i\lambda} a(s) e^{-\int_s^t adp} \Big|_0^t + \frac{1}{i\lambda} \int_0^t e^{i\lambda(t-s)} [a'(s) + a^2(s)] e^{-\int_s^t adp} ds. \tag{4.2.36}$$

Thus

$$J_3 = \frac{a(t)}{-i\lambda} + \frac{e^{i\lambda t - \int_0^t adp} a(0)}{i\lambda} + J_4, \quad \lambda \neq 0, \tag{4.2.37}$$

where J_4 is the last integral in (4.2.36). The first two terms in (4.2.37) tend to zero as $t \to \infty$, due to assumptions (4.2.37), and these assumptions also imply $\lim_{t\to\infty} J_4 = 0$ if $\lambda \neq 0$.

Therefore, passing to the limit $t \to \infty$ in (4.2.33) one gets

$$\lim_{t\to\infty} J_2^2 = ||(E_0 - E_{-0})y||^2 = ||P_{\mathcal{N}} y||^2 = 0. \tag{4.2.38}$$

Theorem 4.2.3 is proved. □

Consider now the case of noisy data f_δ, $||f_\delta - f|| \leq \delta$. Problem (4.2.29), with f_δ in place of f, has a unique solution given by formula (4.2.29) with f_δ in place of f. Let us denote this solution by $u_\delta(t)$. Then

$$||u_\delta(t) - y|| \leq ||u_\delta(t) - u(t)|| + ||u(t) - y|| := I_1 + I_2. \tag{4.2.39}$$

We have

$$\int_0^t e^{-\int_s^t adp} ds \leq \int_0^s \frac{a(s)}{a(t)} e^{-\int_s^t adp} ds \leq \frac{1}{a(t)}.$$

Thus

$$I_1 \leq \delta \int_0^t e^{-\int_s^t adp} ds \leq \frac{\delta}{a(t)}. \tag{4.2.40}$$

In Theorem 4.2.3 we have proved that

$$\lim_{t\to\infty} I_2 = 0. \tag{4.2.41}$$

From (4.2.39) - (4.2.41) we obtain the following result.

Theorem 4.2.4. *Assume (4.2.27) and*

$$\lim_{\delta\to 0} t_\delta = \infty, \quad \lim_{\delta\to 0} \frac{\delta}{a(t_\delta)} = 0. \tag{4.2.42}$$

Define $u_\delta := u_\delta(t_\delta)$, where $u_\delta(t)$ solves (4.2.28) with f_δ replacing f. Then

$$\lim_{\delta\to 0} ||u_\delta - y|| = 0. \tag{4.2.43}$$

Remark 4.2.3. It is of interest to find an optimal in some sense choice of the stopping time t_δ. Conditions (4.2.42) can be satisfied by many choices of t_δ.

Remark 4.2.4. In this Section we assume that A is selfadjoint. If A is not selfadjoint, closed, densely defined in H operator, equation (4.1.1) is solvable, $Ay = f$, $y \perp \mathcal{N}$, then every solution to equation (4.1.1) generates a solution to the equation

$$A^*Au = A^*f \tag{4.2.44}$$

in the following sense. For a densely defined closed linear operator A the operator A^* is also densely defined and closed. However the element f may not belong to the domain $D(A^*)$ of A^*, so equation (4.2.44) has to be interpreted for $f \notin D(A^*)$. We define the solution to (4.2.44) for any $f \in \mathcal{R}(A)$, i.e. for $f = Ay$, $y \perp \mathcal{N}$, by the formula

$$u = \lim_{a \to 0} T_a^{-1} A^* f = y, \quad T = A^*A, \quad T_a := T + aI. \tag{4.2.45}$$

The existence of the limit and the formula

$$\lim_{a \to 0} T_a^{-1} A^* f = y,$$

valid for $y \perp \mathcal{N}$, have been established in Theorem 2.2.1. With the definition (4.2.45) of the solution to (4.2.44) for any $f \in \mathcal{R}(A)$, one has the same equivalence result for the solutions to (4.1.1) and (4.2.44) as in the case of bounded A.

Recall, that if A is bounded and $Au = f$, then $A^*Au = A^*f$, so u which solves (4.1.1) solves (4.2.44) as well. Conversely, if u solves (4.2.44) and $f = Ay$, then $A^*Au = A^*Ay$, so $(A^*A(u-y), u-y) = 0$, and $||Au - Ay|| = 0$, so $Au = f$.

If A is unbounded then we modify the above equivalence claim: if $Ay = f$, $y \perp \mathcal{N}$, then y solves (4.2.44) and vice versa.

With this understanding of the notion of the solution to (4.2.44) we can use Theorems 4.2.1 - 4.2.4 of this Section. Indeed, if equation (4.1.1) is solvable, $Ay = f$, $y \perp \mathcal{N}$, then we replace equation (4.1.1) by equation (4.2.44) with the selfadjoint operator $T = A^*A \geq 0$, and apply Theorems 4.2.1 - 4.2.4 of this Section to equation (4.2.44). Finding the minimal-norm solution to equation (4.1.1) is equivalent, as we have explained, to finding the solution to equation (4.2.44) defined by formula (4.2.45).

4.3 Equations with unbounded operators

The results of Section 4.1 remain valid for unbounded, densely defined, closed linear operator A. In Theorem 2.2.2, we have proved that for such operator A the operator $T_a^{-1}A^*$, $\quad a > 0$, can be considered as defined on all of H bounded operator whose norm is bounded by $\frac{1}{2\sqrt{a}}$ (see (2.2.9)).

Therefore the formulations and proofs of Theorems 4.1.1 - 4.1.3 remain valid.

The results of Section 4.2 have been formulated for unbounded, densely defined, closed linear operators.

In all the results we could assume that the initial condition $u(0)$ is an arbitrary element $u_0 \in H$, not necessarily zero. If u_0 is chosen suitably, for example, in a neighborhood of the solution y, this may accelerate the rate of convergence. *But our results are valid for any choice of* u_0 because

$$\lim_{t\to\infty} ||e^{i(A+ia)t}u_0|| = 0, \quad a > 0,$$

or, in the case $a = a(t)$,

$$\lim_{t\to\infty} ||e^{iAt-\int_0^t a(p)dp}u_0|| = 0.$$

Here the element

$$u_0(t) = e^{iAt-\int_0^t a(p)dp}u_0$$

solves the problem

$$\dot{u} = i[A + ia(t)]u, \quad u(0) = u_0,$$

so that this $u_0(t)$ is a contribution to the solution of (4.2.28) from the non-zero initial condition u_0. If $\int_0^\infty a(t)dt = \infty$, then $\lim_{t\to\infty} u_0(t) = 0$.

4.4 Iterative methods

The main results concerning iterative methods for solving linear ill-posed equation (4.1.1) have been formulated and proved in Section 2.4, Theorems 2.4.1 - 2.4.3.

In this Section we show how to use these results for constructing a stable approximation to the minimal-norm solution y of equation (4.1.1) given noisy data f_δ, $||f_\delta - f|| \leq \delta$.

Our approach is based on a general principle formulated in [R10]:

Proposition 4.4.1. *If equation* $Au = f$ *has a solution and there is a convergent iterative process* $u_{n+1} = T(u_n; f)$, $u_0 = u_0$, *for solving this equation:* $\lim_{n\to\infty} ||u_n - y|| = 0$, *where* $Ay = f$, $y \perp \mathcal{N}$, *and* T *is a continuous operator, then there exists an* $n(\delta)$, $\lim_{\delta\to 0} n(\delta) = \infty$, *such that the same iterative process with* f_δ *replacing* f *produces a sequence* $u_n(f_\delta)$ *such that* $\lim_{\delta\to 0} ||u_{n(\delta)}(f_\delta) - y|| = 0$.

Proof. We have

$$\varepsilon(\delta) := ||u_{n(\delta)}(f_\delta)-y|| \le ||u_{n(\delta)}(f_\delta)-u_{n(\delta)}(f)||+||u_{n(\delta)}(f)-y|| := I_1+I_2. \tag{4.4.1}$$

By one of the assumptions of the Proposition 4.4.1, we have:

$$\lim_{n\to\infty} ||u_n(f) - y|| = 0.$$

For any fixed n, by the continuity of T, we have

$$\lim_{\delta\to 0} ||u_n(f_\delta) - u_n(f)|| := \lim_{\delta\to 0} \eta_n(\delta) = 0. \tag{4.4.2}$$

Thus with

$$w(n) := ||u_n(f) - y||, \quad \lim_{n\to\infty} w(n) = 0,$$

one gets

$$\varepsilon(\delta) \le \eta_{n(\delta)}(\delta) + w(n(\delta)) \to 0 \quad \text{as } \delta \to 0, \tag{4.4.3}$$

where $n(\delta)$ is the minimizer with respect to n of $\eta_n(\delta) + w(n)$ for a small fixed $\delta > 0$.

Proposition 4.4.1 is proved. □

Let us consider iterative process (2.4.1) with f_δ replacing f. Then we have

$$||u_{n+1}(f_\delta) - y|| \le ||u_{n+1}(f_\delta) - u_{n+1}(f)|| + ||u_{n+1}(f) - y||. \tag{4.4.4}$$

It is proved in Theorem 2.4.1 that

$$\lim_{n\to\infty} ||u_n(f) - y|| = 0, \tag{4.4.5}$$

where $Ay = f, \quad y \perp \mathcal{N}$. Furthermore,

$$\varepsilon(n,\delta) := ||u_{n+1}(f_\delta) - u_{n+1}(f)|| \le ||B[u_n(f_\delta) - u_n(f)]|| + \frac{\delta}{2\sqrt{a}}. \tag{4.4.6}$$

Thus

$$\begin{aligned} \varepsilon(n,\delta) &\le \sum_{j=0}^{n} ||B||^j \frac{\delta}{2\sqrt{a}} + ||B^{n+1}|| \; ||u_0(f_\delta) - u_0(f)|| \\ &= \frac{||B||^{n+1} - 1}{||B|| - 1} \frac{\delta}{2\sqrt{a}} := \gamma(n)\delta. \end{aligned}$$

If one denotes

$$w(n) := ||u_n(f) - y||,$$

then

$$||u_{n+1}(f_\delta) - y|| \leq \gamma(n)\delta + w(n) \to 0 \text{ as } \delta \to 0, \qquad \text{if } n = n(\delta), \quad (4.4.7)$$

where $n(\delta)$ is obtained, for example, by minimizing the function

$$\gamma(n)\delta + w(n)$$

with respect to $n = 1, 2, ...$, for a fixed small $\delta > 0$. Note that if $||B|| \geq 1$, then

$$\lim_{n\to\infty} \gamma(n) = \infty.$$

If $||B|| < 1$, then

$$\lim_{n\to\infty} \gamma(n) = \frac{1}{(1 - ||B||)2\sqrt{a}} = \text{const.}$$

In both cases one can take $n(\delta)$ such that $\lim_{\delta\to 0} n(\delta) = \infty$ and

$$\lim_{\delta\to 0} \varepsilon(n(\delta), \delta) = 0. \qquad (4.4.8)$$

We have proved the following result.

Theorem 4.4.2. *Let the assumptions of Theorem 2.4.1 hold, and*

$$||f_\delta - f|| \leq \delta.$$

Then, if one uses the iterative process (2.4.1) with f_δ in place of f and chooses $n(\delta)$ such that (4.4.8) holds, then

$$\lim_{\delta\to 0} ||u_\delta - y|| = 0,$$

where $u_\delta := u_{n(\delta)}(f_\delta)$ is calculated by the above iterative process.

Consider now the iterative process (2.4.8) with f_δ in place of f. An argument, similar to the one given in the proof of Theorem 4.4.2, yields the following result.

Theorem 4.4.3. *Let the assumptions of Theorem 2.4.2 hold, $||f_\delta - f|| \leq \delta$, and $n(\delta)$ is suitably chosen. Then $\lim_{\delta\to 0} ||u_{n(\delta)} - y|| = 0$, where $u_\delta := u_{n(\delta)}$ is calculated by the iterative process (2.4.8) with f_δ in place of f.*

The choice of $n(\delta)$ is quite similar to the choice which was made in the proof of Theorem 4.4.2 above. The role of the operator B is now played by the operator $B = iaA_{ia}^{-1}$ and the role of $\frac{\delta}{2\sqrt{a}}$ is played by $\frac{\delta}{a}$ because

$$||A_{ia}^{-1}|| \leq \frac{1}{a}$$

if $A = A^*$.

Also, using the arguments similar to the ones given in the proof of Theorem 4.4.2 of this Section, one can obtain an analog of Theorem 2.4.3 in the case of noisy data f_δ, $||f_\delta - f|| \leq \delta$.

We leave this to the reader.

4.5 Stable calculation of values of unbounded operators

Let us assume that A is a linear, closed, densely defined operator in H. If $u \in D(A)$ then $Au = f$.

The problem, we study in this Section, is:

How does one calculate f stably given u_δ, $||u_\delta - u|| \leq \delta$?

The element u_δ may not belong to $D(A)$. Therefore this problem is ill-posed.

It was studied in [M] by the variational regularization method.

We propose a new method, an iterative method, for solving the above problem. The equation $Au = f$ is equivalent to the following one:

$$BF = Fu, \tag{4.5.1}$$

where

$$B = (I + Q)^{-1}, \quad Q = AA^*,$$
$$F := (I + Q)^{-1}A = A(I + T)^{-1}, \quad T = A^*A. \tag{4.5.2}$$

It follows (see (2.2.9)) that F is a bounded operator,

$$||F|| \leq \frac{1}{2}, \tag{4.5.3}$$

while B is a selfadjoint operator $0 < B \leq I$, whose eigenspace, corresponding to the eigenvalue $\lambda = 1$, is identical to the null-space $\mathcal{N}(Q)$ of the operator Q, $\mathcal{N}(Q) = \mathcal{N}(A^*) := \mathcal{N}^*$.

If u_δ is given and $||u_\delta - u|| \leq \delta$, then $||Fu_\delta - Fu|| \leq \frac{\delta}{2}$. To solve equation (4.5.1) for f, given the noisy data Fu_δ, we propose to use the following iterative process:

$$f_{n+1} = (I - B)f_n + Fu_\delta := Vf_n + Fu_\delta, \quad V := I - B, \tag{4.5.4}$$

and $f_0 \in H$ is arbitrary. Let g be the minimal-norm solution of equation (4.5.1), i.e., $Bg = Fu, \quad g = Vg + Fu$.

Theorem 4.5.1. *Let $n = n(\delta)$ be an integer such that*

$$\lim_{\delta \to 0} n(\delta) = \infty, \quad \lim_{\delta \to 0} \delta n(\delta) = 0, \tag{4.5.5}$$

and $f_\delta := f_{n(\delta)}$, where f_n is defined by (4.5.4). Then

$$\lim_{\delta \to 0} ||f_\delta - g|| = 0. \tag{4.5.6}$$

Proof. Let $w_n := f_n - g$. Then

$$w_{n+1} = Vw_n + F(u_\delta - u), \quad w_0 = f_0 - g. \tag{4.5.7}$$

From (4.5.7) we derive

$$w_n = \sum_{j=0}^{n-1} V^j F v_\delta + V^n w_0, \quad v_\delta := u_\delta - u, \quad ||v_\delta|| \leq \delta. \tag{4.5.8}$$

Since $||V|| \leq 1$ and $||F|| \leq \frac{1}{2}$, the above equation implies

$$||w_n|| \leq \frac{n\delta}{2} + \left[\int_0^1 (1-s)^{2n} d(E_s w_0, w_0) \right]^{\frac{1}{2}}, \tag{4.5.9}$$

where E_s is the resolution of the identity corresponding to the selfadjoint operator B, $0 < B \leq I$. One has

$$\lim_{n \to \infty} \int_0^1 (1-s)^{2n} d(E_s w_0, w_0) = ||Pw_0||^2 = 0, \tag{4.5.10}$$

where P is the orthoprojector onto the subspace $\mathcal{N}(B)$, but $Pw_0 = 0$ because $\mathcal{N}(B) = \{0\}$. The conclusion (4.5.6) can now be established. Given an arbitrary small $\varepsilon > 0$, find $n = n(\delta)$, sufficiently large so that

$$\int_0^1 (1-s)^{2n} d(E_s w_0, w_0) < \frac{\varepsilon^2}{4}.$$

This is possible as we have already proved (see (4.5.10)).

Simultaneously, we choose $n(\delta)$ so that $\delta n(\delta) < \varepsilon$. This is possible because $\lim_{\delta \to 0} \delta n(\delta) = 0$. Thus, if δ is sufficiently small, then (4.5.6) holds if $n(\delta)$ satisfies (4.5.5).

Theorem 4.5.1 is proved. □

Chapter 5

Some inequalities

In this Chapter some nonlinear inequalities are derived, which are used in other chapters.

5.1 Basic nonlinear differential inequality

We will use much in what follows the following result.

Theorem 5.1.1. *Let $\alpha(t)$, $\beta(t)$, $\gamma(t)$ be continuous nonnegative functions on $[t_0, \infty)$, $t_0 \geq 0$ is a fixed number. If there exists a function*

$$\mu \in C^1[t_0, \infty), \quad \mu > 0, \quad \lim_{t\to\infty} \mu(t) = \infty, \tag{5.1.1}$$

such that

$$0 \leq \alpha(t) \leq \frac{\mu(t)}{2}\left[\gamma(t) - \frac{\dot{\mu}(t)}{\mu(t)}\right], \quad \dot{\mu}(t) = \frac{d\mu}{dt}, \tag{5.1.2}$$

$$\beta(t) \leq \frac{1}{2\mu(t)}\left[\gamma(t) - \frac{\dot{\mu}(t)}{\mu(t)}\right], \tag{5.1.3}$$

$$\mu(0)g(0) < 1, \tag{5.1.4}$$

and $g(t) \geq 0$ satisfies the inequality

$$\dot{g}(t) \leq -\gamma(t)g(t) + \alpha(t)g^2(t) + \beta(t), \quad t \geq t_0, \tag{5.1.5}$$

then $g(t)$ exists on $[t_0, \infty)$ and

$$0 \leq g(t) < \frac{1}{\mu(t)} \to 0 \quad \text{as } t \to \infty. \tag{5.1.6}$$

Remark 5.1.1. There are several novel features in this result. Usually a nonlinear differential inequality is proved by integrating the corresponding differential equation, in our case the following equation:

$$\dot{g}(t) = -\gamma(t)g(t) + \alpha(t)g^2(t) + \beta(t), \tag{5.1.7}$$

and then using a comparison lemma. Equation (5.1.7) is the full Riccati equation the solution to which, in general, may blow up on a finite interval. Assumptions (5.1.2) - (5.1.4) imply that the solutions to (5.1.7) and (5.1.5) exist globally. Furthermore, we do not assume that the coefficient $\alpha(t)$ in front of the senior nonlinear term in (5.1.5) is subordinate to other coefficients.

In fact, in our applications of Theorem 5.1.1 of this Section (see Chapters 6, 7, 10) the coefficient $\alpha(t)$ will tend to infinity as $t \to \infty$.

Remark 5.1.2. Let us give examples of the choices of α, β and γ satisfying assumption (5.1.2) - (5.1.4).

Let

$$\gamma(t) = c_1(1+t)^{\nu_1}, \quad \alpha(t) = c_2(1+t)^{\nu_2}, \quad \beta(t) = c_3(1+t)^{\nu_3},$$

where $c_j > 0$ are constants, and let $\mu = c(1+t)^{\nu}, \quad c = \text{const} > 0, \quad \nu > 0$. If

$$c_2(1+t)^{\nu_2} \le \frac{c(1+t)^{\nu}}{2}\left[c_1(1+t)^{\nu_1} - \frac{1}{1+t}\right],$$

$$c_3(1+t)^{\nu_3} \le \frac{1}{2c(1+t)^{\nu}}\left[c_1(1+t)^{\nu_1} - \frac{1}{1+t}\right], \quad g(0)c < 1,$$

then assumptions (5.1.2) - (5.1.4) hold. The above inequalities hold if, for example:

$$\nu+\nu_1 \ge \nu_2, \quad c_2 \le \frac{c}{2}(c_1-1); \quad \nu_3 \le \nu_1-\nu, \quad c_3 \le \frac{1}{2c}(c_1-1); \quad g(0)c < 1.$$

Let

$$\gamma = \gamma_0 = \text{const} > 0, \quad \alpha(t) = \alpha_0 e^{\nu t}, \quad \beta(t) = \beta_0 e^{-\nu t}, \quad \mu(t) = \mu_0 e^{\nu t},$$

α_0, β_0, μ_0 and ν are positive constants. If

$$\alpha_0 \le \frac{\mu_0}{2}(\gamma_0 - \nu), \quad \beta_0 \le \beta_0 \le \frac{1}{2\mu_0}(\gamma_0 - \nu), \quad g(0)\mu_0 < 1,$$

then assumptions (5.1.2) - (5.1.4) hold.

Let

$$\gamma(t) = [\ln(t + t_1)]^{-\frac{1}{2}}, \quad \mu(t) = c\ln(t + t_1),$$

$$0 \le \alpha(t) \le \frac{c}{2}\left[\sqrt{\ln(t + t_1)} - \frac{1}{t + t_1}\right]$$

and

$$0 \le \beta(t) \le \frac{1}{2c\ln(t + t_1)}\left[\frac{1}{\sqrt{\ln(t + t_1)}} - \frac{1}{(t + t_1)\ln(t + t_1)}\right],$$

where $t_1 = \text{const} > 1$. Then assumptions (5.1.2) - (5.1.4) hold.

Let us recall a known comparison lemma (see, e.g., [H]), whose proof we include for convenience of the reader.

Lemma 5.1.1. *Let $f(t, w)$ and $g(t, u)$ be continuous functions in the region $[0, T) \times D$, $T \le \infty$, $D \subset \mathbb{R}$ is an interval, and $f(t, w) \le g(t, u)$ if $w \le u$, $t \in [0, T)$, $u, w \in D$. Assume that the problem*

$$\dot{u} = g(t, u), \quad u(0) = u_0 \in D,$$

has a unique solution defined on some interval $[0, \tau_u)$, where $\tau_u > 0$. If

$$\dot{w} = f(t, w), \quad w(0) = w_0 \le u_0, \quad w_0 \in D,$$

then $u(t) \ge w(t)$ for all t for which u and w are defined.

Proof of Lemma 5.1.1. Suppose first that $f(t, w) < g(t, u)$ if $w \le u$. Since $w_0 \le u_0$ and $\dot{w}(0) \le f(0, w_0) < g(0, u_0) = \dot{u}(0)$, there exists a $\delta > 0$ such that $u(t) > w(t)$ on $[0, \delta]$. If $u(t_1) \le w(t_1)$ for some $t_1 > \delta$, then there is a $t_2 < t_1$ such that $u(t_2) = w(t_2)$ and $u(t) < w(t)$ for $t \in (t_2, t_1]$. Therefore

$$\dot{w}(t_2) \ge \dot{u}(t_2) = g(t_2, u(t_2)) > f(t_2, w(t_2)) \ge \dot{w}(t_2).$$

This contradiction proves that there is no point t at which $w(t) > u(t)$. Thus, Lemma 5.1.1 is proved under the additional assumption $f(t, w) < g(t, u)$ if $w \le u$.

Let us drop this additional assumption and assume that $f(t, w) \le g(t, u)$ if $w \le u$.

Define $u_n(t)$ as the solution to the problem:

$$\dot{u}_n = g(t, u_n) + \frac{1}{n}, \quad u_n(0) = u_0.$$

Then

$$\dot{w} \le f(t,w) \le g(t,u) < g(t,u) + \frac{1}{n} \quad \text{if} \quad w \le u.$$

By what we have already proved, it follows that

$$w(t) \le u_n(t).$$

Passing to the limit $n \to \infty$, we obtain the conclusion of Lemma 5.1.1. Passing to the limit can be based on the following (also known) result: if $f_j(t,u)$ is a sequence of continuous functions in the region $\mathbb{R}_i := [t_0, t_0 + a] \times |u - u_0| \le \varepsilon$, such that $\lim_{j\to\infty} f_j(t,u) = f(t,u)$ uniformly in $\mathbb{R}$, and

$$\dot{u}_j = f_j(t,u_j), \quad u_j(t_0) = u_0,$$

then

$$\lim_{k\to\infty} u_{j_k}(t) = u(t) \quad \text{uniformly on} \quad [t_0, t - 0 + a],$$

where $\dot{u} = f(t,u)$, $\quad u(t_0) = u_0$; if this problem has a unique solution, then $u_j(t) \to u(t)$ uniformly on $[t_0, t_0 + a]$.

Lemma 5.1.1 is proved. □

Proof of Theorem 5.1.1. Denote $w(t) := g(t)e^{\int_{t_0}^t \gamma(s)ds}$. Then inequality (5.1.5) takes the form

$$\dot{w} \le a(t)w^2 + b(t), \quad w(t_0) = g(0) := g_0, \tag{5.1.8}$$

where

$$a(t) := \alpha(t)e^{-\int_{t_0}^t \gamma(s)ds}, \quad b(t) := \beta(t)e^{\int_{t_0}^t \gamma(s)ds}. \tag{5.1.9}$$

Consider the following Riccati equation

$$\dot{u} = \frac{\dot{f}}{g}u^2(t) - \frac{\dot{g}(t)}{f(t)}. \tag{5.1.10}$$

Its solution can be written analytically:

$$u(t) = -\frac{g(t)}{f(t)} + \frac{1}{f^2(t)\left[c - \int_{t_0}^t \frac{\dot{f}(s)}{g(s)f^2(s)}ds\right]}, \quad c = \text{const}, \tag{5.1.11}$$

(see [Ka]), and one can check that (5.1.11) solves (5.1.10) by direct calculation.

Define

$$f := \mu^{\frac{1}{2}}(t)e^{-\frac{1}{2}\int_{t_0}^t \gamma ds}, \quad g(t) := -\frac{1}{\mu^{\frac{1}{2}}(t)}e^{\frac{1}{2}\int_{t_0}^t \gamma ds}, \tag{5.1.12}$$

and solve the Cauchy problem for equation (5.1.10) with the initial condition

$$u(t_0) = g(t_0). \tag{5.1.13}$$

The solution is given by formula (5.1.11) with

$$c = \frac{1}{\mu(t_0)g(t_0) - 1}. \tag{5.1.14}$$

Let us check assumptions (5.1.2) - (5.1.4). Note that

$$f(t)g(t) = -1, \quad \frac{f}{g} = -\mu(t)e^{-\int_{t_0}^t \gamma ds}, \tag{5.1.15}$$

$$\frac{\dot{f}}{g} = \frac{\mu}{2}\left(\gamma - \frac{\dot{\mu}}{\mu}\right)e^{-\int_{t_0}^t \gamma ds} \geq a(t) \geq 0, \tag{5.1.16}$$

$$-\frac{\dot{g}}{f} = \frac{\left(\gamma - \frac{\dot{\mu}}{\mu}\right)}{2\mu}e^{\int_{t_0}^t \gamma ds} \geq b(t) \geq 0, \tag{5.1.17}$$

where conditions (5.1.2) - (5.1.3) imply (5.1.16) and (5.1.17). Condition (5.1.4) implies

$$c < 0. \tag{5.1.18}$$

Let us apply Lemma 5.1.1 to problems (5.1.8) and (5.1.10). The assumptions of this Lemma are satisfied due to (5.1.13), (5.1.16) and (5.1.17). Thus, Lemma 5.1.1 implies

$$w(t) \leq u(t), \quad t \geq t_0, \tag{5.1.19}$$

that is

$$g(t)e^{\int_{t_0}^t \gamma ds} \leq \frac{e^{\int_{t_0}^t \gamma ds}}{\mu(t)}\left[1 - \frac{1}{\frac{1}{1-\mu(t_0)g(t_0)} + \frac{1}{2}\int_{t_0}^t \left(\gamma - \frac{\dot{\mu}}{\mu}\right)ds}\right]. \tag{5.1.20}$$

This inequality implies

$$g(t) \leq \frac{1}{\mu(t)}, \tag{5.1.21}$$

because $\gamma - \frac{\dot{\mu}}{\mu} \geq 0$.

Theorem 5.1.1 is proved. □

5.2 An operator inequality

In this Section we state and prove an operator version of the widely used Gronwall inequality for scalar functions. This inequality can be formulated as follows:

If $a(t), g(t), b(t) \geq 0$ are continuous functions, and

$$g(t) \leq b(t) + \int_0^t a(s)g(s)ds, \quad 0 \leq t \leq T,$$

then

$$g(t) \leq b(t) + \int_0^t e^{\int_s^t a(p)dp} b(s)ds, \quad 0 \leq t \leq T. \tag{5.2.1}$$

Its proof is simple and can be found in many books and in Section 5.4 below.

A nonlinear version of this result is:

If $u(t, g)$ is a continuous function, nondecreasing with respect to g in the region $R := [t_0, t_0 + a] \times \mathbb{R}$, and v solves the inequality

$$v(t) \leq v_0 + \int_{t_0}^t u(s, v(s))ds, \quad v_0 \leq g_0,$$

then

$$v(t) \leq g(t), \quad t \in [t_0, t_0 + a],$$

where $g(t)$ is the maximal solution of the problem

$$\dot{y} = u(t, g), \quad g(t_0) = g_0,$$

and we assume that v and g exist on $[t_0, t_0 + a]$.

Several Gronwall-type inequalities for scalar functions are proved in Section 5.4. In this Section we prove an operator version of the Gronwall inequality, which we will use in Chapter 10. The result is stated in the following Theorem.

Theorem 5.2.1. *Let $Q(t)$, $T(t)$ and $G(t)$ be bounded linear operator functions in a Hilbert space defined for $t \geq 0$, and assume that*

$$\dot{Q} = -T(t)Q(t) + G(t), \quad Q(0) = Q_0, \quad \dot{Q} := \frac{dQ}{dt}, \tag{5.2.2}$$

where the derivative can be understood in a weak sense. Assume that there exists a positive integrable function $\varepsilon(t)$ such that

$$(T(t)h, h) \geq \varepsilon(t)||h||^2, \quad \forall h \in H. \tag{5.2.3}$$

Then

$$||Q(t)|| \le e^{-\int_0^t \varepsilon(s)ds} \left[||Q_0|| + \int_0^t ||G(s)|| e^{-\int_0^s \varepsilon(p)dp} ds \right], \tag{5.2.4}$$

where $Q_0 := Q(0)$.

Proof. Denote

$$g(t) = ||Q(t)h||,$$

where $h \in H$ is arbitrary. Equation (5.2.2) implies

$$g\dot{g} = \mathrm{Re}(\dot{Q}h, Qh) = -(TQh, Qh) + \mathrm{Re}(Gh, Qh) \le -\varepsilon(t)g^2 + ||Gh||g.$$

Since $g \ge 0$, we get

$$\dot{g} \le -\varepsilon(t)g + ||G(t)h||. \tag{5.2.5}$$

This inequality implies

$$g(t) \le g(0)e^{-\int_0^t \varepsilon(s)ds} + \int_0^t ||G(s)h|| e^{-\int_s^t \varepsilon(p)dp} ds. \tag{5.2.6}$$

Taking supremum in (5.2.6) with respect to all h on the unit sphere, $||h|| = 1$, we obtain inequality (5.2.4).

Theorem 5.2.1 is proved. □

5.3 A nonlinear inequality

Let

$$\dot{u} \le -a(t)f(u(t)) + b(t), \quad u(0) = u_0, \quad u \ge 0. \tag{5.3.1}$$

Assume that $f(u) \in \mathrm{Lip}_{loc}[0, \infty)$, $a(t), b(t) \ge 0$ are continuous functions on $[0, \infty)$,

$$\int_0^\infty a(s)ds = \infty, \quad \lim_{t\to\infty} \frac{b(t)}{a(t)} = 0, \quad a \ge 0, \quad b \ge 0, \tag{5.3.2}$$

$$f(0) = 0; \quad f(u) > 0 \text{ for } u > 0; \quad f(u) \ge c > 0 \text{ for } u \ge 1, \tag{5.3.3}$$

where c is a positive constant.

Theorem 5.3.1. *These assumptions imply global existence of* $u(t)$ *and its decay at infinity:*

$$\lim_{t\to\infty} u(t) = 0. \tag{5.3.4}$$

Proof of Theorem 5.3.1. Assumptions (5.3.2) about $a(t)$ allow one to introduce the new variable

$$s = s(t) := \int_0^t a(p)dp,$$

and claim that the map $t \to s$ maps $[0, \infty)$ onto $[0, \infty)$. In the new variable inequality (5.3.1) takes the form

$$w' \leq -f(w) + \beta(s), \quad w(0) = u_0, \quad w \geq 0, \tag{5.3.5}$$

where

$$w = w(s) = u(t(s)), \quad w' = \frac{dw}{ds}, \quad \text{and} \quad \beta = \frac{b(t(s))}{a(t(s))}, \quad \lim_{s\to\infty} \beta(s) = 0.$$

Since

$$f(u) \in \text{Lip}_{loc}[0, \infty),$$

there exists locally and is unique the solution to the Cauchy problem

$$\dot{v} = -f(v) + \beta(s), \quad v(0) = u_0.$$

Let us prove that (5.3.5) and assumptions (5.3.3) about f imply global existence of v, and, therefore, of w, and its decay at infinity:

$$\lim_{s\to\infty} w(s) = 0. \tag{5.3.6}$$

The global existence of v follows from a bound $||v(s)|| \leq c$ with c independent of s (see Lemma 2.6.1). This bound is proved as a similar bound for w, which follows from (5.3.6). If (5.3.6) is established, then Theorem 5.3.1 is proved.

To prove (5.3.6) define the set $E \subset \mathbb{R}_+ := [0, \infty)$:

$$E := \{s : f(w(s)) \leq \beta(s) + \frac{1}{1+s}\}, \tag{5.3.7}$$

and let

$$F := \mathbb{R}_+ \backslash E.$$

Let us prove that

$$\sup E = +\infty. \tag{5.3.8}$$

If $\sup E := k < \infty$, then

$$f(w(s)) - \beta > \frac{1}{1+s}, \qquad s > k.$$

This and inequality (5.3.5) imply

$$w' \leq -\frac{1}{1+s}, \qquad s > k. \tag{5.3.9}$$

Integrating (5.3.9) from k to ∞, one gets

$$\lim_{s \to \infty} w(s) = -\infty,$$

which contradicts the assumption $w \geq 0$. Thus (5.3.8) is established.

Let us derive (5.3.6) from (5.3.8). Let $s_1 \in E$ be such a point that $(s_1, s_2) \in F, \quad s_2 > s_1$. Then

$$f(w(s)) > \beta(s) + \frac{1}{1+s}, \qquad s \in (s_1, s_2), \tag{5.3.10}$$

so

$$w' \leq -\frac{1}{1+s}, \qquad s \in (s_1, s_2). \tag{5.3.11}$$

Integrating, we get

$$w(s) - w(s_1) \leq -\ln\left(\frac{1+s}{1+s_1}\right) < 0.$$

Thus

$$w(s) < w(s_1), \qquad s \in (s_1, s_2). \tag{5.3.12}$$

However, $s_1 \in E$, so

$$f(w(s_1)) \leq \frac{1}{s_1+1} + \beta(s_1) \to 0 \qquad \text{as } s_1 \to \infty. \tag{5.3.13}$$

From (5.3.8), (5.3.12) and (5.3.13) the relation (5.3.6) follows, and the conclusion (5.3.4) of Theorem 5.3.1 is established.

Theorem 5.3.1 is proved. □

Let us now prove the following result.

Theorem 5.3.2. *Assume that $y(t)$ and $h(t)$ are nonnegative continuous functions, defined on $[0, \infty)$, and*

$$\int_0^\infty [h(t) + y(t)]dt < \infty,$$

and suppose that the following inequality holds:

$$y(t) - y(s) \le \int_s^t f(y(p))dp + \int_s^t h(p)dp, \tag{5.3.14}$$

where $f > 0$ is a nondecreasing continuous function on $[0, \infty)$. Then

$$\lim_{t \to \infty} y(t) = 0. \tag{5.3.15}$$

Proof of Theorem 5.3.2 Assume the contrary: there exists $t_n \to \infty$ such that $y(t_n) \ge a > 0$. Choose t_{n+1} such that

$$t_{n+1} - t_n \ge \frac{a}{2f(a)} := c. \tag{5.3.16}$$

Let

$$m_n := \min_{t_n - c \le p \le t_n} y(p) := y(p_n), \quad p_n \in [t_n - c, t_n]. \tag{5.3.17}$$

Denote

$$\lambda_n := \sup\{t : p_n < t < t_n, \quad y(t) < a\}. \tag{5.3.18}$$

By the continuity of $y(t)$, we have $y(\lambda_n) = a$. Therefore

$$y(\lambda_n) - y(p_n) = a - m_n \le \int_{p_n}^{\lambda_n} f(y(s))ds + \int_{p_n}^{\lambda_n} h(s)ds \le cf(a) + \delta_n, \tag{5.3.19}$$

where $\delta_n := \int_{p_n}^{\lambda_n} h(s)ds$, and we took into account that $p_n \in [t_n - c, t_n]$ and $\lambda_n \in [p_n, t_n]$, so that $\lambda_n - p_n \le c = \frac{a}{2f(a)}$. From (5.3.19) we derive

$$\frac{a}{2} \le m_n + \delta_n \le y(t) + \delta_n, \quad t \in [t_n - c, t_n]. \tag{5.3.20}$$

Integrate (5.3.20) over $[t_n - c, t_n]$ and get:

$$\frac{ac}{2} \le \int_{t_n - c}^{t_n} y(s)ds + c\delta_n. \tag{5.3.21}$$

Sum up (5.3.21) from $n = 1$ to $n = N$ to get

$$\frac{ac}{2}N \le \sum_{j=1}^{N}\int_{t_j-c}^{t_j} y(s)ds + c\sum_{j=1}^{N}\delta_j. \tag{5.3.22}$$

Let $N \to \infty$. By assumption, $\int_0^\infty [y(s) + h(s)]ds < \infty$, so that the right-hand side of (5.3.22) is bounded as $N \to \infty$, while its left-hand side is not.

This contradiction proves Theorem 5.3.2. □

Remark 5.3.1. Theorem 5.3.2 is proved in [ZS], p. 227, by a different argument. In [ZS] several applications of such result are given to nonlinear partial differential equations.

5.4 The Gronwall-type inequalities

Let

$$\dot{v} \le a(t)v + b(t), \quad t \ge t_0, \tag{5.4.1}$$

where $a(t)$ and $b(t)$ are integrable functions. Then the Gronwall inequality is:

$$v(t) \le v(t_0)e^{\int_{t_0}^t a(s)ds} + \int_{t_0}^t b(s)e^{\int_s^t a(p)dp}ds, \quad t \ge t_0. \tag{5.4.2}$$

To prove it, multiply (5.4.1) by $e^{-\int_{t_0}^t a(s)ds}$. Then (5.4.1) yields

$$\frac{d}{dt}[e^{-\int_{t_0}^t a(s)ds}v(t)] \le b(t)e^{-\int_{t_0}^t a(s)ds}. \tag{5.4.3}$$

Thus

$$e^{-\int_{t_0}^t a(s)ds}v(t) \le v(t_0) + \int_{t_0}^t b(s)e^{-\int_{t_0}^s a(p)dp}ds. \tag{5.4.4}$$

This is equivalent to (5.4.2).

Lemma 5.4.1. *Assume now that*

$$v, a, b \ge 0, \tag{5.4.5}$$

$$\int_t^{t+r} a(s)ds \le a_1, \quad \int_t^{t+r} b(s)ds \le a_2, \quad \int_t^{t+r} v(s)ds \le a_3, \tag{5.4.6}$$

where $r \ge 0$. Then (5.4.1) implies

$$v(t+r) \le (a_2 + \frac{a_3}{r})e^{a_1}, \quad t \ge t_0, \quad r \ge 0. \tag{5.4.7}$$

Proof. To prove (5.4.7), use (5.4.3) and (5.4.5) to get

$$\frac{d}{dt}[e^{-\int_{t_0}^{t} adp} v(t)] \leq b(t). \tag{5.4.8}$$

From (5.4.8) we obtain

$$e^{-\int_{t_0}^{t+r} adp} v(t+r) \leq e^{-\int_{t_0}^{t_1} adp} v(t_1) + \int_{t_1}^{t+r} b(p)dp, \quad t_0 \leq t_1 < t+r, \tag{5.4.9}$$

and

$$v(t+r) \leq (v(t_1) + a_2)e^{a_1}. \tag{5.4.10}$$

Integrating (5.4.10) with respect to t_1 over the interval $[t, t+r]$ yields (5.4.7).

Lemma 5.4.1 is proved. □

Inequality (5.4.7) is proved, e.g., in [Te].

Lemma 5.4.2. *Suppose c and M are positive constants,*

$$u_j \leq c + M \sum_{m=0}^{j-1} u_m, \quad u_0 \leq c, \quad u_m \in \mathbb{R}. \tag{5.4.11}$$

Then

$$u_p \leq c(1+M)^p, \quad p = 1, 2, ... \tag{5.4.12}$$

Proof. Inequality (5.4.12) is easy to prove by induction: for $p = 0$ inequality (5.4.12) holds due to (5.4.11). If (5.4.12) holds for $p \leq n$, then it holds for $p_m = n + 1$ due to (5.4.11).

$$u_{n+1} \leq M \sum_{m=0}^{n} c(1+M)^m \leq c(1 + M\frac{(1+M)^{n+1} - 1}{1 + M - 1}) = c(1+M)^{n+1}. \tag{5.4.13}$$

Thus, (5.4.12) is proved.

Lemma 5.4.2 is proved. □

Chapter 6

DSM for monotone operators

In this Chapter the DSM method is developed for solving equations with monotone operators. Convergence of the DSM is proved for any initial approximation. The DSM yields the unique minimal-norm solution.

6.1 Auxiliary results

In this Chapter we study the equation

$$F(u) = f, \tag{6.1.1}$$

assuming that F is monotone in the sense

$$(F(u) - F(v), u - v) \geq 0, \quad \forall u, v \in H, \tag{6.1.2}$$

where H is a Hilbert space. We assume also that $F \in C^2_{\text{loc}}$, i.e. assumptions (1.3.2) hold, but we do not assume that assumption (1.3.1) hold. Therefore, the problem of solving (6.1.1) cannot be solved, in general, by Newton's method. We assume that equation (6.1.1) has a solution, possibly non-unique. For $F \in C^1_{\text{loc}}$, condition (6.1.2) is equivalent to

$$F'(u) \geq 0. \tag{6.1.3}$$

We will use the notations:

$$A := F'(u), \quad A_a := A + aI. \tag{6.1.4}$$

In Section 6.2 we formulate a result which contains a justification of a DSM for solving equation (6.1.1) with exact data f, and then we show how to use this DSM for a stable approximation of the solution y given noisy data f_δ, $||f_\delta - f|| \leq \delta$. In this Section we prove some auxiliary results used in the next Section.

Let us recall some (known) properties of monotone operators. For convenience of the reader we prove the results we use in this Chapter. These results are auxiliary for the rest of the Chapter.

Lemma 6.1.1. *Equation (6.1.1) holds if and only if*

$$(F(v) - f, v - u) \geq 0 \quad \forall v \in H. \tag{6.1.5}$$

Proof of Lemma 6.1.1. If u solves (6.1.1) then (6.1.5) holds by the monotonicity (6.1.2). Conversely, if (6.1.5) holds, take $v = u + \lambda w, \quad \lambda = \text{const} > 0, \quad w \in H$ is arbitrary, then (6.1.5) yields

$$(F(u + \lambda w) - f, w) \geq 0. \tag{6.1.6}$$

Let $\lambda \to 0$ and use the continuity of F to get

$$(F(u) - f, w) \geq 0. \tag{6.1.7}$$

Since w is arbitrary, this implies (6.1.1). Lemma 6.1.1 is proved. □

Lemma 6.1.2. *The set* $\mathcal{N} := \{u : F(u) - f = 0\}$ *is closed and convex.*

Proof of Lemma 6.1.2. The set $\mathcal{N}$ is closed if F is continuous. Let us prove that $\mathcal{N}$ is convex. Assume that $u, w \in \mathcal{N}$. We want to prove that $\lambda u + (1 - \lambda)w \in \mathcal{N}$ for any $\lambda \in (0, 1)$. By Lemma 6.1.1 we have to prove:

$$0 \leq (F(v) - f, v - \lambda u - (1 - \lambda)w) = \lambda(F(v) - f, v - u) + (1 - \lambda)(F(v) - f, u - w),$$

so that the desired inequality follows from the assumption $u, w \in \mathcal{N}$ and from Lemma 6.1.1.

Lemma 6.1.2 is proved. □

Lemma 6.1.3. *A closed and convex set* $\mathcal{N}$ *in a Hilbert space has a unique element of minimal norm.*

Proof of Lemma 6.1.3. Assume that

$$||u|| \leq ||v|| \quad \forall v \in \mathcal{N}$$

and

$$||w|| \leq ||v||, \quad \forall v \in \mathcal{N}.$$

Since $\mathcal{N}$ is convex, we have

$$||u|| \leq ||\frac{u + w}{2}||, \quad ||w|| \leq ||\frac{u + w}{2}||, \quad ||u|| = ||w||.$$

Thus $(u, w) = ||u||||w||$, so $u = v$. Lemma 6.1.3 is proved. □

Remark 6.1.1. A Banach space X is called strictly convex if the condition

$$||u|| = ||w|| = ||\frac{u+w}{2}||$$

implies $u = w$. Hilbert spaces are strictly convex, $L^p(a, b)$ space for $p > 1$ is strictly convex, but $C([a, b])$ and $L^1([a, b])$ are not strictly convex. If X is strictly convex, then $||u + w|| = ||u|| + ||w||$ implies $w = \lambda u$, where λ is a real number.

Lemma 6.1.4. *(w-closedness). If (6.1.2) holds and F is continuous, then the assumptions*

$$u_n \rightharpoonup u, \quad F(u_n) \to f \tag{6.1.8}$$

imply (6.1.1). Here $\rightharpoonup$ denotes weak convergence.

Proof of Lemma 6.1.4. Using (6.1.2) we have

$$(F(v) - F(u_n), v - u_n) \geq 0. \tag{6.1.9}$$

Passing to the limit $n \to \infty$ in (6.1.9) and using (6.1.8) we obtain

$$(F(v) - f, u - v) \geq 0 \quad \forall v \in H. \tag{6.1.10}$$

By Lemma 6.1.1, this implies $F(u) = f$.
Lemma 6.1.4 is proved. □

Lemma 6.1.5. *If (6.1.2) holds and $F \in C^1_{loc}$, then*

$$||A_a^{-1}|| \leq \frac{1}{a}, \quad a = const > 0; \quad A := F'(u). \tag{6.1.11}$$

Proof of Lemma 6.1.5. By (6.1.3) one has $A \geq 0$. Thus,

$$(A_a v, v) \geq a||v||^2, \quad \forall v \in H.$$

Since $(Av, v) \leq ||Av|| \, ||v||$, one gets $||A_a v|| \geq a||v||$. Thus, with $A_a v := w$, one has

$$a^{-1}||w|| \geq ||A_a^{-1} w||.$$

This implies (6.1.11).
Lemma 6.1.5 is proved. □

Lemma 6.1.6. *Equation*

$$F(u) + au = f, \quad a = const > 0 \tag{6.1.12}$$

has a unique solution for every $f \in H$ if $F \in C^2_{loc}$ satisfies (6.1.2).

Proof of Lemma 6.1.6. Consider the problem

$$\dot{u} = -A_a^{-1}[F(u) + au - f], \quad u(0) = u_0, \tag{6.1.13}$$

where $u_0 \in H$ is arbitrary and A_a is defined in (6.1.4). Since $F \in C^2_{\text{loc}}$, the operator in the right-hand side of (6.1.13) satisfies locally a Lipschitz condition, so problem (6.1.13) has a unique local solution $u(t)$. Define

$$g(t) := ||F(u(t)) + au(t) - f||.$$

Then, by equation (6.1.13), one gets

$$\dot{g} = -g, \quad g(0) = ||F(u_0)) + au_0 - f|| := g_0; \quad g(t) \leq g_0 e^{-t}. \tag{6.1.14}$$

Using (6.1.13) again and applying (6.1.11) we get

$$||\dot{u}|| \leq \frac{g_0}{a} e^{-t},$$

so

$$\int_0^\infty ||\dot{u}|| dt \leq \frac{g_0}{a},$$

there exists $u(\infty)$, and

$$||u(t) - u_0|| \leq \frac{g_0}{a}, \quad ||u(\infty) - u(t)|| \leq \frac{g_0}{a} e^{-t}. \tag{6.1.15}$$

From (6.1.11) and (6.1.15) we conclude that $u(\infty)$ solves equation (6.1.12). The solution to this equation is unique: if u and v solve (6.1.12), then

$$F(u) - F(v) + a(u - v) = 0.$$

Multiplying this equation by $u - v$ and using (6.1.2), we derive $u = v$.

Lemma 6.1.6 is proved. □

Remark 6.1.2. The assumption $F \in C^2_{\text{loc}}$ in Lemma 6.1.6 can be relaxed: F is hemicontinuous , defined on all of H, suffices for the conclusion of Lemma 6.1.6 to hold. ([V]).

Lemma 6.1.7. *Assume that (6.1.11) is solvable, y is its minimal-norm solution, and assumptions (6.1.2) and (1.3.2) hold. Then*

$$\lim_{a \to 0} ||u_a - y|| = 0, \tag{6.1.16}$$

where u_a solves (6.1.12).

Proof of Lemma 6.1.7. We note that $F(y) = f$ and write equation (6.1.12) as

$$F(u_a) - F(y) + au_a = 0, \quad a > 0. \tag{6.1.17}$$

Multiply this equation by $u_a - y$, use (6.1.12) and get:

$$(u_a, u_a - y) \le 0. \tag{6.1.18}$$

Thus

$$||u_a|| \le ||y||. \tag{6.1.19}$$

Inequality (6.1.19) implies the existence of a sequence $u_n := u_{a_n}$, $a_n \to 0$ as $n \to \infty$, such that

$$u_n \rightharpoonup u. \tag{6.1.20}$$

From equation (6.1.12) it follows that

$$F(u_n) \to f. \tag{6.1.21}$$

From (6.1.20), (6.1.21) and Lemma 6.1.4 it follows that

$$F(u) = f. \tag{6.1.22}$$

Let us prove that $u = y$. Indeed, (6.1.19) implies

$$\overline{\lim_{n\to\infty}} ||u_n|| \le ||y||, \tag{6.1.23}$$

while (6.1.20) implies

$$||u|| \le \underline{\lim}_{n\to\infty} ||u_n||. \tag{6.1.24}$$

Combine (6.1.23) and (6.1.24) to get

$$||u|| \le ||y||. \tag{6.1.25}$$

Since u and y are solutions to equation (6.1.1) and y is the minimal-norm solution, which is unique by Lemmas 6.1.1 and 6.1.2, it follows that $u = y$. Therefore (6.1.20) implies

$$u \rightharpoonup y, \tag{6.1.26}$$

while (6.1.23) and (6.1.24) imply

$$\lim_{n\to\infty} ||u_n|| = ||y||. \tag{6.1.27}$$

From (6.1.26) and (6.1.27) it follows that

$$\lim_{n\to\infty} ||u_n - y|| = 0. \tag{6.1.28}$$

Indeed,

$$||u_n - y||^2 = ||u_n||^2 + ||y||^2 - 2\mathrm{Re}(u_n, y) \underset{n\to\infty}{\longrightarrow} 0. \tag{6.1.29}$$

Therefore every convergent sequence $u_n = u_{a_n}$ converges strongly to y. This implies (6.1.16).

Lemma 6.1.7 is proved. □

Remark 6.1.3. One can estimate the rate $||\dot{V}||$ of convergence of $V(t)$ to y when $V(t)$ solves equation (6.1.12) with $a = a(t)$. We assume

$$0 < a(t) \searrow 0, \quad \lim_{t\to\infty} \frac{\dot{a}(t)}{a(t)} = 0, \quad \frac{|\dot{a}(t)|}{a(t)} \le \frac{1}{2}. \tag{6.1.30}$$

For example one may take

$$a(t) = \frac{c_1}{(c_0 + t)^b},$$

where $c_0, c_1, b > 0$ are constants, and $2b \le c_0$.

Differentiating equation (6.1.12) with respect to t, one gets:

$$A_{a(t)}\dot{V} = -\dot{a}V. \tag{6.1.31}$$

Thus

$$||\dot{V}|| \le |\dot{a}| \; ||A_{a(t)}^{-1}V|| \le \frac{|\dot{a}(t)|}{a(t)}||V|| \le \frac{|\dot{a}(t)|}{a(t)}||y||. \tag{6.1.32}$$

Lemma 6.1.8. *If*

$$\lim_{t\to\infty} a(t) = 0, \quad a(t) > 0,$$

then

$$\lim_{t\to\infty} ||V(t) - y|| = 0. \tag{6.1.33}$$

Proof of Lemma 6.1.8. The conclusion (6.1.33) follows immediately from (6.1.16) and the assumption $\lim_{t\to\infty} a(t) = 0$.

If one assumes more about $a(t)$ then one still can not estimate the rate of convergence in (6.1.33), see also Remark 6.1.4 below.

Lemma 6.1.8 is proved. □

Remark 6.1.4. It is not possible to use estimate (6.1.32) in order to estimate $||V(t) - y||$ because such an estimate requires the integral $\int_t^\infty \frac{|\dot{a}(s)|}{a(s)} ds$ to converge. However, this integral diverges for any $a(t)$ satisfying (6.1.30). Indeed $|\dot{a}| = -\dot{a}$, so

$$\int_t^\infty \frac{|\dot{a}(s)|}{a(s)} ds = -\lim_{N\to\infty} \int_t^N \frac{\dot{a}}{a} ds = -\lim_{N\to\infty} \ln \frac{a(N)}{a(t)} = \lim_{N\to\infty} \ln \frac{a(t)}{a(N)} = \infty,$$

because $\lim_{N\to\infty} a(N) = 0$.

Therefore, without extra assumptions about f in equation (6.1.1) it is not possible to estimate the rate of convergence in (6.1.33).

6.2 Formulation of the results and proofs

Theorem 6.2.1. *Consider the DSM*

$$\dot{u} = -A_{a(t)}^{-1}(u)[F(u) + a(t)u - f], \quad u(0) = u_0, \tag{6.2.1}$$

where $u_0 \in H$ is arbitrary, F satisfies (6.1.2) and (1.3.2), and $a(t)$ satisfies (6.1.30). Assume that equation (6.1.2) has a solution. Then problem (6.2.1) has a unique global solution $u(t)$, there exists $u(\infty)$, and $u(\infty)$ solves equation (6.1.1).

Proof of Theorem 6.2.1. Denote

$$w := u(t) - V(t),$$

where $V(t)$ solves equation (6.1.12) with $a = a(t)$. Then

$$||u(t) - y|| \le ||w|| + ||V(t) - y||. \tag{6.2.2}$$

We have proved (6.1.33). We want to prove

$$\lim_{t\to\infty} ||w(t)|| = 0. \tag{6.2.3}$$

We derive a differential inequality (5.1.5) for $g(t) = ||w(t)||$ and then use Theorem 5.1.1.

Let us proceed according to this plan. Write equation (6.2.1) as

$$\dot{w} = -\dot{V} - A_{a(t)}^{-1}[F(u) - F(V) + a(t)w]. \tag{6.2.4}$$

We use Taylor's formula and get:

$$F(u) - F(V) + aw = A_a(u)w + \varepsilon, \quad ||\varepsilon|| \le \frac{M_2}{2}||w||^2, \tag{6.2.5}$$

where M_2 is the constant from the estimate

$$\sup_{u\in B(u_0,R)} ||F''(u)|| \leq M_2(R) := M_2.$$

Denote

$$g(t) := ||w(t)||.$$

Multiply (6.2.4) by w, and using (6.2.5), get:

$$g\dot{g} \leq -g^2 + \frac{M_2}{2}||A_{a(t)}^{-1}||g^3 + ||\dot{V}||g. \tag{6.2.6}$$

Since $g \geq 0$, we get the inequality

$$\dot{g} \leq -g(t) + \frac{c_0}{a(t)}g^2 + \frac{|\dot{a}|}{a(t)}c_1, \quad c_0 := \frac{M_2}{2}, \quad c_1 := ||y||, \tag{6.2.7}$$

where we have used the estimates (6.1.11) and (6.1.32). Inequality (6.2.7) is of the type (5.1.5) with

$$\gamma(t) := 1, \quad \alpha(t) := \frac{c_0}{a(t)}, \quad \beta(t) := c_1\frac{|\dot{a}|}{a(t)}. \tag{6.2.8}$$

Let us check assumptions (5.1.2) - (5.1.4). We take

$$\mu(t) = \lambda a(t), \quad \lambda = \text{const}.$$

Assumption (5.1.2) is:

$$\frac{c_0}{a(t)} \leq \frac{\lambda}{2a(t)}\left[1 - \frac{|\dot{a}|}{a(t)}\right], \tag{6.2.9}$$

where we took into account that $|\dot{a}| = -\dot{a}$. Since

$$\frac{|\dot{a}|}{a} \leq \frac{1}{2},$$

assumption (6.2.9) will be satisfied if the following inequality holds:

$$c_0 \leq \frac{\lambda}{4}. \tag{6.2.10}$$

Take $\lambda \geq 4c_0$ and (6.2.10) is satisfied. Assumption (5.1.3) is:

$$c_1\frac{|\dot{a}|}{a(t)} \leq \frac{a(t)}{2\lambda}\left[1 - \frac{|\dot{a}|}{a}\right]. \tag{6.2.11}$$

This assumption holds if

$$4\lambda c_1 \frac{|\dot{a}(t)|}{a^2(t)} \leq 1. \tag{6.2.12}$$

The scaling transformation

$$a(t) \to \nu a(t),$$

where $\nu = \text{const} > 0$, does not change assumptions (6.1.30), but makes the ratio $\frac{|\dot{a}|}{a^2}$ as small as one wishes if ν is sufficiently large. So taking $\nu a(t)$ in place of $a(t)$ one can satisfy inequality (6.2.12). Finally, assumption (5.1.4) is

$$\frac{\lambda}{a(0)} g(0) < 1. \tag{6.2.13}$$

This assumption is satisfied for any fixed $g(0)$ and λ if $a(0)$ is sufficiently large. Again, taking $\nu a(t)$ in place of $a(t)$ one can obtain arbitrarily large $\nu a(0)$ and satisfy inequality (6.2.13) (with $a(0)$ replaced by $\nu a(0)$ with a sufficiently large constant $\nu > 0$).

Thus, Theorem 5.1.1 yields:

$$g(t) < \frac{a(t)}{\lambda} \to 0 \quad \text{as } t \to \infty. \tag{6.2.14}$$

This implies uniform with respect to $t \in [0, \infty)$ boundedness of the norm $||u(t)||$ of the unique local solution to (6.2.1) and therefore existence of its unique global solution, existence of the limit $u(\infty)$ and the relation $u(\infty) = y$, where y is the (unique) minimal-norm solution to equation (6.1.1).

Theorem 6.2.1 is proved. □

Theorem 6.2.2. *Assume that* $a = \text{const} > 0$, *and*

$$\dot{u} = -A_a^{-1}[F(u) + au - f], \quad u(0) = u_0, \tag{6.2.15}$$

where $u_0 \in H$ *is arbitrary, (6.1.2) and (1.3.2) hold. Assume that equation (6.1.1) is solvable and* y *is its minimal-norm solution. Then problem (6.2.15) has a unique global solution* $u_a(t)$ *and*

$$\lim_{a \to 0} \lim_{t \to \infty} ||u_a(t) - y|| = 0. \tag{6.2.16}$$

Proof of Theorem 6.2.2. Local existence of the unique solution to (6.2.15) follows from the local Lipschitz condition satisfied by the operator in the right-hand side of (6.2.15). Global existence of this solution follows from the uniform boundedness with respect to $t \to \infty$ of the norm $||u_a(t)||$ of the solution to (6.2.15). This boundedness is established by the same argument as in the proof of Lemma 6.1.6. Also, as in this proof one establishes the existence of $u_a(\infty) = \lim_{t\to\infty} u_a(t)$, and the relation

$$F(u_a(\infty)) + au(\infty) = f.$$

Finally, this and Lemma 6.1.7 from Section 6.1 imply (6.2.16).

Theorem 6.2.2 is proved. □

Remark 6.2.1. If we integrate problem (6.2.15) over an interval of length τ, then we can choose $a(\tau)$ such that

$$\lim_{\tau\to\infty} a(\tau) = 0$$

and

$$\lim_{\tau\to\infty} ||u_{a(\tau)}(\tau) - y|| = 0. \tag{6.2.17}$$

This observation allows us to choose $a = a(\tau)$ if the length τ of the interval of integration is chosen a priori and then increased. Consequently one calculates the solution y using one limiting process in equation (6.2.17) rather than two limiting processes in equation (6.2.16).

6.3 The case of noisy data

Let us use Theorem 6.2.1 in the case when f is replaced by f_δ in equation (6.2.1). Denote by u_δ the solution of (6.2.1) with f_δ replacing f. Then

$$||u_\delta(t) - y|| \leq ||u_\delta(t) - V_\delta(t)|| + ||V_\delta(t) - V(t)|| + ||V(t) - y||, \tag{6.3.1}$$

where $V(t)$ solves equation (6.1.12) and V_δ solves equation (6.1.2) with f_δ in place of f.

We have already proved (6.1.33). We want to prove that if $t = t_\delta$ in (6.3.1), where t_δ, the stopping time, is properly chosen, then

$$\lim_{\delta\to 0} ||u_\delta(t_\delta) - V_\delta(t_\delta)|| = 0, \quad \lim_{\delta\to 0} ||V_\delta(t_\delta) - V(t_\delta)|| = 0. \tag{6.3.2}$$

Lemma 6.3.1. *One has*

$$||V_\delta(t) - V(t)|| \leq \frac{\delta}{a(t)}. \tag{6.3.3}$$

Thus,

$$\lim_{\delta\to 0} ||V_\delta(t_\delta) - V(t_\delta)|| = 0,$$

if

$$\lim_{\delta\to 0} \frac{\delta}{a(t_\delta)} = 0. \tag{6.3.4}$$

Proof of Lemma 6.3.1. From equation (6.1.12) we derive

$$F(V_\delta) - F(V) + a(t)(V_\delta - V) - (f_\delta - f) = 0. \tag{6.3.5}$$

Multiplying (6.3.5) by $V_\delta - V$, using (6.1.2) and the inequality $||f_\delta - f|| \le \delta$, we obtain:

$$a(t)||V_\delta - V||^2 \le \delta ||V_\delta - V||.$$

This implies (6.3.3). Lemma 6.3.1 is proved. □

Let us estimate $||u_\delta(t) - V_\delta(t)|| := g(t)$. We use the ideas of the proof of Theorem 6.2.1. Let $w := u_\delta - V_\delta(t)$. Then

$$\dot{w} = -\dot{V}_\delta - A^{-1}_{a(t)}[F(u_\delta) - F(V_\delta) + a(t)w],$$

which is an equation similar to (6.2.4). As in the proof of Theorem 6.2.1, we obtain

$$\dot{g} \le -g(t) + \frac{c_0}{a(t)} g^2 + \frac{|\dot{a}|}{a(t)} c_1, \tag{6.3.6}$$

and conclude that the estimate similar to(6.2.14) holds:

$$g(t) \le \frac{a(t)}{\lambda}. \tag{6.3.7}$$

Therefore, if t_δ is such that

$$\lim_{\delta\to 0} t_\delta = \infty, \quad \lim_{\delta\to 0} \frac{\delta}{a(t_\delta)} = 0, \tag{6.3.8}$$

then (6.3.2) holds and

$$\lim_{\delta\to 0} ||u_\delta(t_\delta) - y|| = 0. \tag{6.3.9}$$

We have proved the following result.

Theorem 6.3.1. *Assume that $||f_\delta - f|| \le \delta$, t_δ satisfies (6.3.8), and $u_\delta(t)$ solves problem (6.2.1) with f_δ in place of f. Then (6.3.9)holds.*

Remark 6.3.1. The results of this Chapter can be generalized: the condition

$$\mathrm{Re}(F(u) - F(v), u - v) \ge 0 \tag{6.3.10}$$

can be used in place of (6.1.2).

Chapter 7

DSM for general nonlinear operator equations

In this Chapter we construct a convergent DSM for solving any solvable nonlinear equation $F(u) = 0$, under the assumptions $F \in C^2_{\text{loc}}, \quad F'(y) \neq 0$, where $F(y) = 0$.

7.1 Formulation of the problem. The results and proofs

Consider the equation

$$F(u) = 0, \tag{7.1.1}$$

where $F : H \to H$ satisfies assumption (1.3.2), equation (7.1.1) has a solution y, possibly non-unique, and

$$\tilde{A} := F'(y) \neq 0. \tag{7.1.2}$$

Under these assumptions we want to construct a DSM method for solving equation (7.1.1).

Consider the following DSM method:

$$\dot{u} = -T^{-1}_{a(t)}[A^*F(u) + a(t)(u - z)], \quad u(0) = u_0, \tag{7.1.3}$$

where $u_0 \in H$, $z \in H$. We use the following notations:

$$A := F'(u), \quad T := A^*A, \quad T_a := T + aI, \quad \tilde{T} := \tilde{A}^*\tilde{A}, \quad \tilde{A} := F'(y). \tag{7.1.4}$$

If assumption (7.1.2) holds, then one can find an element $v \neq 0$ such that

$$\tilde{T}v \neq 0.$$

Thus, there is an element z such that

$$y - z = \tilde{T}v,$$

and one can choose z such that $||v|| \ll 1$, i.e. $||v|| > 0$ is as small as one wishes:

$$y - z = \tilde{T}v, \quad ||v|| \ll 1. \tag{7.1.5}$$

Since y is unknown, we do not give an algorithm for finding such a z, but only prove its existence.

Let us denote

$$w := u(t) - y.$$

Since $F(y) = 0$, one has, using the Taylor's formula:

$$F(u) = F(u) - F(y) = Aw + \varepsilon, \quad ||\varepsilon|| \leq \frac{M_2}{2}||w||^2.$$

Here and below $M_j, j = 1, 2$, are constants from (1.3.2). Equation (7.1.3) can be written as:

$$\dot{w} = -T_{a(t)}^{-1}[(A^*A + a(t))w + A^*\varepsilon + a(t)\tilde{T}v].$$

Let

$$g(t) := ||w(t)||.$$

Multiplying the above equation by w we get:

$$g\dot{g} \leq -g^2 + \frac{M_2 g^3}{4\sqrt{a(t)}} + a(t)||T_{a(t)}^{-1}\tilde{T}||\ ||v||g, \tag{7.1.6}$$

where we have used estimate (7.1.4) and the inequality (2.1.13):

$$||T_a^{-1}A^*|| \leq \frac{1}{2\sqrt{a(t)}}. \tag{7.1.7}$$

Since $g \geq 0$, we get

$$\dot{g} \leq -g + \frac{c_0 g^2}{\sqrt{a(t)}} + a(t)||v||\ ||T_{a(t)}^{-1}\tilde{T}||, \qquad c_0 := \frac{M_2}{4}. \tag{7.1.8}$$

Let us transform the last factor:

$$||T_{a(t)}^{-1}\tilde{T}|| \leq || \left(T_{a(t)}^{-1} - \tilde{T}_{a(t)}^{-1}\right) \tilde{T}|| + ||\tilde{T}_{a(t)}^{-1}\tilde{T}||. \tag{7.1.9}$$

We have

$$||\tilde{T}_{a(t)}^{-1}\tilde{T}|| \leq 1, \quad ||aT_a^{-1}|| \leq 1,$$

and

$$T_a^{-1} - \tilde{T}_a^{-1} = T_a^{-1}\left(\tilde{T} - T\right)\tilde{T}_a^{-1}. \tag{7.1.10}$$

Moreover

$$||\tilde{T} - T|| = ||\tilde{A}^*\tilde{A} - A^*A|| \leq 2M_1M_2||u - y|| = 2M_1M_2 g. \tag{7.1.11}$$

Therefore, choosing $||v||$ so that

$$2M_1M_2||v|| \leq \frac{1}{2}, \tag{7.1.12}$$

one can rewrite (7.1.8) as

$$\dot{g} \leq -\frac{1}{2}g(t) + \frac{c_0 g^2(t)}{\sqrt{a(t)}} + a(t)||v||, \quad t \geq 0, \quad c_0 := \frac{M_2}{4}. \tag{7.1.13}$$

Let

$$\mu = \frac{\lambda}{\sqrt{a(t)}}. \tag{7.1.14}$$

Let us apply Theorem 5.1.1 to inequality (7.1.13). We have

$$\gamma(t) = \frac{1}{2}, \quad \alpha(t) = \frac{c_0}{\sqrt{a(t)}}, \quad \beta(t) = a(t)||v||.$$

Condition (5.1.2) is:

$$\frac{c_0}{\sqrt{a(t)}} \leq \frac{\lambda}{2\sqrt{a(t)}}\left(\frac{1}{2} + \frac{\dot{a}}{2a}\right) = \frac{\lambda}{4\sqrt{a(t)}}\left(1 - \frac{|\dot{a}|}{a}\right). \tag{7.1.15}$$

Let us assume that

$$0 < a(t) \searrow 0, \quad \frac{|\dot{a}(t)|}{a} \leq \frac{1}{2}. \tag{7.1.16}$$

For example, one can take

$$a(t) = \frac{c_1}{(c_2+t)^b},$$

where $c_1,\ c_2,\ b > 0$ are constants, $2b \le c_2$. Inequality (7.1.16) is satisfied for rapidly decaying $a(t)$ as well, e.g.,

$$a(t) = e^{-ct}, \quad c \le \frac{1}{2}.$$

Inequality (7.1.15) holds if

$$\lambda \ge 8c_0. \tag{7.1.17}$$

Condition (5.1.3) is:

$$a(t)||v|| \le \frac{\sqrt{a(t)}}{2\lambda}\left(1 - \frac{|\dot{a}|}{a}\right). \tag{7.1.18}$$

Because of inequality (7.1.16), this inequality holds if

$$4\lambda\sqrt{a(0)}||v|| \le 1. \tag{7.1.19}$$

If

$$||v|| \le \frac{1}{4\lambda\sqrt{a(0)}}, \tag{7.1.20}$$

then (7.1.19) holds.

Finally, condition (5.1.4) is:

$$\frac{\lambda}{\sqrt{a(0)}} g(0) < 1. \tag{7.1.21}$$

This condition holds if

$$||u_0 - y|| < \frac{\sqrt{a(0)}}{\lambda}. \tag{7.1.22}$$

Inequality (7.1.22) is valid for any initial approximation u_0 provided that λ is sufficiently large. For any fixed λ inequality (7.1.20) holds if $||v||$ is sufficiently small. Therefore, if conditions (7.1.16), (7.1.17), (7.1.20) and (7.1.22) hold, then (5.1.6) implies:

$$||u(t) - y|| < \frac{\sqrt{a(t)}}{\lambda} \to 0 \text{ as } t \to \infty. \tag{7.1.23}$$

Let us formulate the result we have just proved.

Theorem 7.1.1. *Assume that equation (7.1.1) has a solution* y, $F(y) = 0$, *possibly non-unique, and the following conditions are satisfied: (1.3.2), (7.1.2), (7.1.5), (7.1.16), (7.1.17), (7.1.20) and (7.1.22). Then problem (7.1.3) has a unique solution* $u(t)$, *there exists* $u(\infty)$, $u(\infty) = y$, *and the rate of convergence of* $u(t)$ *to the solution* y *is given by (7.1.23).*

7.2 Noisy data

Assume now that the noisy data f_δ are given,

$$||f_\delta - f|| \leq \delta,$$

equation (7.1.1) is of the form

$$F(u) = f, \tag{7.2.1}$$

and this equation has a solution y, $F(y) = f$.

Consider the DSM similar to (7.1.3):

$$\dot{u}_\delta = -T^{-1}_{a(t)}[A^*(F(u_\delta) - f_\delta) + a(t)(u_\delta - z)], \quad u_\delta(0) = u_0. \tag{7.2.2}$$

As in Section 7.1, denote

$$w = w_\delta := u_\delta - y,$$

and by

$$g := g_\delta := ||w_\delta(t)||.$$

We want to prove that for a suitable stopping time t_δ, $\lim_{\delta \to 0} t_\delta = \infty$, the quantity $w_\delta(t_\delta) \to 0$ as $\delta \to 0$, in other words,

$$\lim_{\delta \to 0} ||u_\delta - y|| = 0, \quad u_\delta := u_\delta(t_\delta). \tag{7.2.3}$$

Let us argue as in Section 7.1. We have now

$$f_\delta = f + \eta_\delta, \quad ||\eta_\delta|| \leq \delta.$$

Therefore, inequality (7.1.13) will be replaced by

$$\dot{g} \leq -\frac{1}{2}g(t) + \frac{c_0 g^2(t)}{\sqrt{a(t)}} + a(t)||v|| + \frac{\delta}{2\sqrt{a(t)}}, \qquad c_0 := \frac{M_2}{4}, \tag{7.2.4}$$

where (7.1.12) holds. Let us choose

$$\mu(t) = \frac{\lambda}{\sqrt{a(t)}}$$

as in (7.1.14), and assume (7.1.16), (7.1.17) and (7.1.20). Then conditions (5.1.2) and (5.1.4) are satisfied. Condition (5.1.3) is satisfied if

$$a(t)||v|| + \frac{\delta}{2\sqrt{a(t)}} \leq \frac{\sqrt{a(t)}}{4\lambda}. \tag{7.2.5}$$

Let us choose v small enough, so that

$$a(t)||v|| \leq \frac{1}{2}\frac{\sqrt{a(t)}}{4\lambda},$$

that is,

$$4\lambda\sqrt{a(0)}||v|| \leq \frac{1}{2}, \tag{7.2.6}$$

and t_δ such that

$$\frac{\delta}{2\sqrt{a(t)}} \leq \frac{1}{2}\frac{\sqrt{a(t)}}{4\lambda},$$

that is,

$$\frac{2\lambda\delta}{a(t)} \leq \frac{1}{2}, \qquad t \leq t_\delta. \tag{7.2.7}$$

Then (7.2.5) holds for $t \leq t_\delta$, and estimate (5.1.6) yields

$$||u_\delta - y|| \leq \frac{\sqrt{a(t)}}{\lambda}, \quad t \leq t_\delta. \tag{7.2.8}$$

Therefore, if t_δ is chosen as the solution to the equation

$$a(t) = 4\lambda\delta, \tag{7.2.9}$$

then $u_\delta := u_\delta(t_\delta)$ satisfies the estimate

$$||u_\delta - y|| \leq 2\sqrt{\frac{\delta}{\lambda}}. \tag{7.2.10}$$

We have proved the following result.

Theorem 7.2.1. *Assume equation (7.2.1) has a solution y, possibly non-unique, $||f_\delta - f|| \leq \delta$, and the following conditions are satisfied: (1.3.2), (7.1.2), (7.1.3), (7.1.16), (7.1.17), (7.1.22), (7.2.6) and t_δ satisfies (7.2.9). Then problem (7.2.2) has a unique solution $u_\delta(t)$ on the interval $[0, t_\delta]$, and $u_\delta := u_\delta(t_\delta)$ satisfies inequality (7.2.10).*

Remark 7.2.1. Theorem 7.2.1 shows that, in principle, DSM (7.2.2) can be used for solving stably any solvable operator equation (7.2.1) for which $F \in C^2_{\text{loc}}$, i.e. assumption (1.3.2) holds, and F satisfies assumption (7.1.2), where y is a solution to equation (7.2.1).

These are rather weak assumptions. However, as it was mentioned already, we do not give an algorithmic choice of z in (7.2.2). Under an additional assumption, e.g., if one assumes $y = \tilde{T}v, \quad ||v|| \ll 1$, it is possible to take $z = 0$ and, using the arguments given in the proofs of Theorems 7.1.1 and 7.2.1, establish the conclusions of these theorems.

7.3 Iterative solution

Let us prove the following result.

Theorem 7.3.1. *Under the assumptions of Theorem 7.1.1, the iterative process*

$$u_{n+1} = u_n - h_n T_{a_n}^{-1}[A^*(u_n)F(u_n) + a_n(u_n - z)], \quad u_0 = u_0, \quad (7.3.1)$$

where $h_n > 0$ and $a_n > 0$ are suitably chosen, generates the sequence u_n converging to y.

Remark 7.3.1. The suitable choices of a_n and h_n are discussed in the proof of Theorem 7.3.1.

Lemma 7.3.1. *Let*

$$g_{n+1} \le \gamma g_n + p g_n^2, \quad g_0 := m > 0, \quad 0 < \gamma < 1, \quad p > 0.$$

If

$$m < \frac{q - \gamma}{p}, \quad \text{where} \quad \gamma < q < 1,$$

then

$$\lim_{n\to\infty} g_n = 0, \quad \text{and} \quad g_n \le g_0 q^n.$$

Proof. Estimate

$$g_1 \le \gamma m + p m^2 \le q m$$

holds if

$$m \le \frac{q - \gamma}{p}, \quad \gamma < q < 1.$$

Assume that $g_n \le g_0 q^n$. Then

$$g_{n+1} \le \gamma g_0 q^n + p(g_0 q^n)^2 = g_0 q^n(\gamma + p g_0 q^n) < g_0 q^{n+1},$$

because

$$\gamma + p g_0 q^n < \gamma + p g_0 q \le q.$$

Lemma 7.3.1 is proved. □

Proof of Theorem 7.3.1.
Let

$$w_n := u_n - y, \quad g_n := \|w_n\|.$$

As in the proof of Theorem 7.1.1, we assume

$$2M_1 M_2 \|v\| \le \frac{1}{2}$$

and rewrite (7.3.1) as

$$\begin{aligned} w_{n+1} &= w_n - h_n\, T_{a_n}^{-1}\, [A^*(u_n)(F(u_n) - F(y)) + a_n w_n + a_n(y - z)], \\ w_0 &= \|u_0 - y\|. \end{aligned}$$

Using the Taylor formula

$$F(u_n) - F(y) = A(u_n) w_n + K(w_n), \quad \|K\| \le \frac{M_2 g_n^2}{2},$$

the estimate

$$\|T_{a_n}^{-1} A^*(u_n)\| \le \frac{1}{2\sqrt{a_n}},$$

and the formula

$$y - z = \tilde{T} v,$$

we get

$$w_{n+1} = (1 - h_n) w_n - h_n T_{a_n}^{-1} A^*(u_n) K(w_n) - h_n a_n T_{a_n}^{-1} \tilde{T} v. \qquad (7.3.2)$$

Taking into account that

$$\|\tilde{T}_a^{-1} \tilde{T}\| \le 1,$$

and

$$a\|T_a^{-1}\| \leq 1 \quad \text{if} \quad a > 0,$$

we obtain

$$\|T_{a_n}^{-1}\tilde{T}v\| \leq \|(T_{a_n}^{-1} - \tilde{T}_{a_n}^{-1})\tilde{T}\|\|v\| + \|v\|,$$

and

$$\|(T_{a_n}^{-1} - \tilde{T}_{a_n}^{-1})\tilde{T}\| = \|T_{a_n}^{-1}(\tilde{T}_{a_n} - T_{a_n})\tilde{T}_{a_n}^{-1}\tilde{T}\| \leq \frac{2M_1M_2g_n}{a_n} := \frac{c_1g_n}{a_n}.$$

Let

$$c_0 := \frac{M_2}{4}.$$

Then we obtain from (7.3.2) the following inequality:

$$g_{n+1} \leq (1 - h_n)g_n + \frac{c_0h_ng_n^2}{\sqrt{a_n}} + c_1h_n\|v\|g_n + h_na_n\|v\|.$$

We have assumed in the proof of Theorem 7.3.1 that

$$c_1\|v\| \leq \frac{1}{2}.$$

Thus

$$g_{n+1} \leq (1 - \frac{h_n}{2})g_n + \frac{c_0h_n}{\sqrt{a_n}}g_n^2 + h_na_n\|v\|.$$

Choose

$$a_n = 16c_0^2g_n^2.$$

Then $\frac{c_0g_n}{\sqrt{a_n}} = \frac{1}{4}$, and

$$g_{n+1} \leq (1 - \frac{h_n}{4})g_n + 16c_0^2h_n\|v\|g_n^2, \qquad g_0 = \|u_0 - y\| \leq R, \quad (7.3.3)$$

where $R > 0$ is defined in (1.3.2). Take $h_n = h \in (0, 1)$ and choose $g_0 := m$ such that

$$m < \frac{q + h - 1}{16c_0h\|v\|},$$

where $q \in (0, 1)$ and $q + h > 1$.

Then Lemma 7.3.1 implies

$$\|u_n - y\| \leq g_0q^n \to 0 \text{ as } n \to \infty.$$

Theorem 7.3.1 is proved. □

7.4 Stability of the iterative solution

Assume that the equation we want to solve is $F(u) = f$, f is unknown, f_δ is known, $||f_\delta - f|| \leq \delta$. Consider the iterative process similar to (7.3.1):

$$v_{n+1} = v_n - h_n T_{a_n}^{-1}[A^*(v_n)(F(v_n) - f_\delta) + a_n(v_n - z)], \quad v_0 = u_0, \quad (7.4.1)$$

Let

$$w_n := v_n - y, \quad ||w_n|| := \psi_n,$$

and choose $h_n = h$ independent of n, $h \in (0,1)$. Later we impose a positive lower bound on h, see formula (7.4.6) below. The inequality similar to (7.3.3) is:

$$\psi_{n+1} \leq \gamma\psi_n + p\psi_n^2 + \frac{h\delta}{2\sqrt{a_n}}, \quad \psi_0 = ||u_0 - y||, \tag{7.4.2}$$

where

$$\gamma := 1 - \frac{h}{4}, \quad p := 16c_0^2||v||, \quad a_n = 16c_0^2\psi_n^2. \tag{7.4.3}$$

We stop iterations in (7.4.2) when $n = n(\delta)$, where $n(\delta)$ is the largest integer for which the following inequality

$$\frac{h\delta}{2\sqrt{a_n}} \leq \kappa\gamma\psi_n, \quad \kappa \in \left(0, \frac{1}{3}\right)$$

holds. One can use formula (7.4.3) for a_n and rewrite this inequality in the form:

$$\frac{h\delta}{8c_0\kappa\gamma} \leq \psi_n^2, \quad \kappa \in (0, \frac{1}{3}). \tag{7.4.4}$$

If (7.4.4) holds, then (7.4.2) implies

$$\psi_{n+1} \leq (1+\kappa)\gamma\psi_n + p\psi_n^2, \quad (1+\kappa)\gamma < 1, \tag{7.4.5}$$

where the conditions

$$\gamma = 1 - \frac{h}{4}, \quad 0 < \kappa < \frac{1}{3}, \quad h \in \left(\frac{4\kappa}{1+\kappa}, 1\right) \tag{7.4.6}$$

imply that $(1+\kappa)\gamma < 1$. If

$$\psi_0 < \frac{q - (1+\kappa)\gamma}{p}, \quad \text{where } (1+\kappa)\gamma < q < 1, \quad \gamma = 1 - \frac{h}{4}, \tag{7.4.7}$$

then (7.4.5) and Lemma 7.3.1 imply

$$\psi_n \leq \psi_0 q^n, \qquad \text{for } n < n(\delta), \quad (1+\kappa)\gamma < q < 1, \quad 0 < \kappa < \frac{1}{3}, \tag{7.4.8}$$

where $n(\delta)$ is the largest integer for which inequality (7.4.4) holds. We have

$$\lim_{\delta \to 0} n(\delta) = \infty.$$

Thus

$$\lim_{\delta \to 0} \psi_{n(\delta)} = 0. \tag{7.4.9}$$

We have proved the following result.

Theorem 7.4.1. *Let the assumptions of Theorem 7.1.1 and conditions (7.4.4), (7.4.6), and (7.4.7) hold, and* ψ_n *be defined by (7.4.1). Then relations (7.4.8) and (7.4.9) hold, so* $\lim_{\delta \to 0} ||v_{n(\delta)} - y|| = 0$.

Chapter 8

DSM for operators satisfying a spectral assumption

In this Chapter we introduce a spectral assumption (8.1.1) and obtain some results based on this assumption.

8.1 Spectral assumption

In this Chapter we assume that the operator equation (7.2.1) is solvable, y is its solution, possibly non-unique, $F \in C^2_{\text{loc}}$, and the linear operator $A = F'(u)$ has the set $\{z : |\arg z - \pi| \leq \varphi_0, \quad 0 < |z| < r_0\}$ consisting of regular points of F', where $\varphi_0 > 0$ and $r_0 > 0$ are arbitrary small fixed numbers, and $u \in H$ is arbitrary. This is a *spectral assumption* on F. If this condition is satisfied then

$$||(A + \varepsilon I)^{-1}|| \leq \frac{c_0}{\varepsilon}, \quad 0 < \varepsilon \leq r_0, \tag{8.1.1}$$

where $c_0 = \frac{1}{\sin \varphi_0}$.

Condition (8.1.1) is much weaker than the assumption of monotonicity of F. If F is monotone, then $A = F'(u) \geq 0$ and $||A_\varepsilon^{-1}|| \leq \frac{1}{\varepsilon}$ for all $\varepsilon > 0$, where $A_\varepsilon = A + \varepsilon I$. If $c_0 = 1$ and $r_0 = \infty$ in (8.1.1), then A is a generator of a semigroup of contractions ([P]).

It is known that if F is monotone, hemicontinuous, and $\varepsilon > 0$, then the equation

$$F(u) + \varepsilon u = f, \quad \varepsilon = \text{const} > 0, \tag{8.1.2}$$

is uniquely solvable for any $f \in H$. We want to prove a similar result assuming (8.1.1).

Our first result is the following.

Theorem 8.1.1. *Assume that F satisfies conditions (1.3.2) and (8.1.1). Then equation (8.1.2) has a solution.*

Proof of Theorem 8.1.1. From our assumptions it follows that the problem

$$\dot{u} = -A_{\varepsilon}^{-1}[F(u) + \varepsilon u - f], \quad u(0) = u_0, \tag{8.1.3}$$

is locally solvable. We will prove the uniform bound

$$\sup_{t \geq 0} ||u(t)|| \leq c, \tag{8.1.4}$$

where $c = \text{const} > 0$ does not depend on t and the supremum is over all t for which $u(t)$, the local solution to (8.1.3), does exist. If (8.1.4) holds, then, as we have proved in Section 2.6, Lemma 2.6.1, the local solution is a global one, it exists on $[0, \infty)$. We also prove that there exists $u(\infty)$, and that $u(\infty)$ solves (8.1.2).

Let us prove estimate (8.1.4). Consider the function

$$g(t) := ||F(u(t)) + \varepsilon u(t) - f||. \tag{8.1.5}$$

We have

$$g\dot{g} = \text{Re}([F'(u(t)) + \varepsilon]\dot{u}, F(u) + \varepsilon u - f) = -g^2. \tag{8.1.6}$$

Thus

$$g(t) = g_0 e^{-t}, \quad g_0 := g(0). \tag{8.1.7}$$

From (8.1.1), (8.1.3) and (8.1.7) we get

$$||\dot{u}|| \leq \frac{c_0 g_0 e^{-t}}{\varepsilon}. \tag{8.1.8}$$

This implies the existence of $u(\infty)$ and the estimates

$$||u(t) - u(0)|| \leq \frac{c_0 g_0}{\varepsilon}, \quad ||u(t) - u(\infty)|| \leq \frac{c_0 g_0 e^{-t}}{\varepsilon}. \tag{8.1.9}$$

For any fixed u_0 and $\varepsilon > 0$, one can take $R > 0$ such that

$$\frac{c_0 g_0}{\varepsilon} \leq R,$$

so that the trajectory $u(t)$ stays in the ball $B(u_0, R)$. Passing to the limit $t \to \infty$ in (8.1.7) yields

$$F(u(\infty)) + \varepsilon u(\infty) - f = 0. \tag{8.1.10}$$

Therefore $u(\infty)$ solves equation (8.1.2), so this equation is solvable. Moreover, the DSM (8.1.3) converges to a solution $u(\infty)$ at an exponential rate, see (8.1.9).

Theorem 8.1.1 is proved. □

Remark 8.1.1. Estimate (8.1.1) follows from the spectral assumption because of the known estimate of the norm of the resolvent of a linear operator. Namely, if A is a linear operator and $(A - zI)^{-1}$ is its resolvent, where z is a complex number in the set of regular points of the operator A. A point z is called a regular point of A if the operator $A - zI$ has a bounded inverse defined on the whole space. Otherwise z is called a point of spectrum σ of A.

The estimate we have mentioned is:

$$||(A - zI)^{-1}|| < \frac{1}{\rho(z, \sigma)}, \tag{8.1.11}$$

where $\rho(z, \sigma)$ is the distance (on the complex plane $\mathbb{C}$) from the point z to the set σ of points of spectrum of A. To check this, consider the function

$$(A - zI - \mu I)^{-1} = (A - zI)^{-1}[I - \mu(A - zI)^{-1}]^{-1}.$$

If

$$|\mu| \; ||(A - zI)^{-1}|| < 1,$$

then the above function is an analytic function of μ. Therefore, if μ is smaller than the distance from z to the nearest point of spectrum of A, then the point $z + \mu$ is a regular point of A.

Thus (8.1.11) follows.

Remark 8.1.2. We have proved in Theorem 8.1.1 that for $\varepsilon \in (0, r_0)$ equation (8.1.2) is solvable. This does not imply that the limiting equation

$$F(u) = f, \tag{8.1.12}$$

is solvable.

For example, the equation $e^u + \varepsilon u = 0$ is solvable in $\mathbb{R}$ for any $\varepsilon > 0$, but the limiting equation $e^u = 0$ has no solutions.

Therefore it is of interest to study the following problem:

If the limiting equation (8.1.12) is solvable, then what are the conditions under which the solution to equation (8.1.2), or a more general equation

$$F(u_\varepsilon) + \varepsilon(u_\varepsilon - z) = f, \tag{8.1.13}$$

where $z \in H$ is some element, converges to a solution to (8.1.12) as $\varepsilon \to 0$?

This question is discussed in Chapter 9.

8.2 Existence of a solution to a nonlinear equation

Let $D \subset \mathbb{R}^3$ be a bounded domain with Lipschitz boundary S, $k = const > 0$, and $f : \mathbb{R} \to \mathbb{R}$ be a function such that

$$uf(u) \geq 0, \quad \text{for} \quad |u| \geq a \geq 0, \tag{8.2.1}$$

where a is an arbitrary nonnegative fixed number. We assume that f is continuous in the region $|u| \geq a$, and bounded and piecewise-continuous with at most finitely many discontinuity points u_j, such that $f(u_j+0)$ and $f(u_j-0)$ exist, in the region $|u| \leq a$.

Consider the problem

$$(-\Delta + k^2)u + f(u) = 0 \quad \text{in} \quad D, \tag{8.2.2}$$

$$u = 0 \quad \text{on} \quad S. \tag{8.2.3}$$

There is a large literature on problems of this type. Usually it is assumed that f does not grow too fast or f is monotone (see, e.g., [B] and references therein).

The novel point in this Section is the absence of the monotonicity restrictions on f and of the growth restrictions on f as $|u| \to \infty$, except for the assumption (8.2.1).

This assumption allows an arbitrary behavior of f inside the region $|u| \leq a$, where $a \geq 0$ can be arbitrary large, and an arbitrarily rapid growth of f to $+\infty$ as $u \to +\infty$, or arbitrarily rapid decay of f to $-\infty$ as $u \to -\infty$.

Our result is:

Theorem 8.2.1. *Under the above assumptions problem* (8.2.2)–(8.2.3) *has a solution* $u \in H^2(D) \cap \overset{\circ}{H}{}^1(D) := H_0^2(D)$.

Here $H^\ell(D)$ is the usual Sobolev space, $\overset{\circ}{H}{}^1(D)$ is the closure of $C_0^\infty(D)$ in the norm $H^1(D)$. Uniqueness of the solution does not hold without extra assumptions.

The ideas of our proof are: first, we prove that if

$$\sup_{u \in \mathbb{R}} |f(u)| \leq \mu,$$

then a solution to (8.2.2)–(8.2.3) exists by the Schauder's fixed-point theorem (see Section 16, Theorem 16.8.1). Here μ is a constant. Secondly, we prove an a priori bound

$$\|u\|_\infty \leq a.$$

If this bound is proved, then problem (8.2.2)–(8.2.3) with the nonlinearity f replaced by

$$F(u) := \begin{cases} f(u), & |u| \leq a \\ f(a), & u \geq a \\ f(-a), & u \leq -a \end{cases} \tag{8.2.4}$$

has a solution, and this solution solves the original problem (8.2.2)–(8.2.3). The bound

$$\|u\|_\infty \leq a$$

is proved by using some integral inequalities. An alternative proof of this bound is also given. This proof is based on the maximum principle for elliptic equation (8.2.2).

We use some ideas from [R9]. Our presentation follows [R56].

Proof of Theorem 8.2.1. If $u \in L^\infty := L^\infty(D)$, then problem (8.2.2)–(8.2.3) is equivalent to the integral equation:

$$u = -\int_D G(x,y) f(u(y)) dy := T(u), \tag{8.2.5}$$

where

$$(-\Delta + k^2)G = -\delta(x-y) \quad \text{in} \quad D, \qquad g\,|_{x\in S} = 0. \tag{8.2.6}$$

By the maximum principle,

$$0 \leq G(x,y) < g(x,y) := \frac{e^{-k|x-y|}}{4\pi|x-y|}, \qquad x, y \in D. \tag{8.2.7}$$

The map T is a continuous and compact map in the space $C(D) := X$, and

$$\|u\|_{C(D)} := \|u\| \leq \mu \sup_x \int_D \frac{e^{-k|x-y|}}{4\pi|x-y|} dy \leq \mu \int_{\mathbb{R}^3} \frac{e^{-k|y|}}{4\pi|y|} dy \leq \frac{\mu}{k^2}. \tag{8.2.8}$$

This is an a priori estimate of any bounded solution to (8.2.2)–(8.2.3) for a bounded nonlinearity f such that

$$\sup_{u\in\mathbb{R}} |f(u)| \leq \mu.$$

Thus, Schauder's fixed-point theorem yields the existence of a solution to (8.2.5), and consequently to problem (8.2.2)–(8.2.3), for bounded f. Indeed, if B is a closed ball of radius $\frac{\mu}{k^2}$, then the map T maps this ball into itself by (8.2.8), and since T is compact, the Schauder principle is applicable. Thus, the following lemma is proved.

Lemma 8.2.1. *If* $\sup_{u\in\mathbb{R}}|f(u)| \le \mu$, *then problems* (8.2.5) *and* (8.2.2)–(8.2.3) *have a solution in* $C(D)$, *and this solution satisfies estimate* (8.2.8).

Let us now prove an a priori bound for any solution $u \in C(D)$ of the problem (8.2.2)–(8.2.3) without assuming that $\sup_{u\in\mathbb{R}}|f(u)| < \infty$.

Let

$$u_+ := \max(u, 0).$$

Multiply (8.2.2) by $(u-a)_+$, integrate over D and then by parts to get

$$0 = \int_D [\nabla u \cdot \nabla (u-a)_+ + k^2 u(u-a)_+ + f(u)(u-a)_+]dx, \qquad (8.2.9)$$

where the boundary integral vanishes because

$$(u-a)_+ = 0 \quad \text{on} \quad S \quad \text{for} \quad a \ge 0.$$

Each of the terms in (8.2.9) is nonnegative, the last one due to (8.2.1). Thus (8.2.9) implies

$$u \le a. \qquad (8.2.10)$$

Similarly, using (8.2.1) again, and multiplying (8.2.2) by $(-u-a)_+$, one gets

$$-a \le u. \qquad (8.2.11)$$

We have proved:

Lemma 8.2.2. *If* (8.2.1) *holds, then any solution* $u \in H_0^2(D)$ *to* (8.2.2)–(8.2.3) *satisfies the inequality*

$$|u(x)| \le a. \qquad (8.2.12)$$

Consider now equation (8.2.5) in $C(D)$ with an arbitrary continuous f satisfying (8.2.1). Any $u \in C(D)$ which solves (8.2.5), solves (8.2.2)-(8.2.3), and therefore satisfies (8.2.12) and belongs to $H_0^2(D)$. This u solves problem (8.2.2)–(8.2.3) with f replaced by F, defined in (8.2.4), and vice versa. Since F is a bounded nonlinearity, equation (8.2.5) and problem (8.2.2)–(8.2.3) (with f replaced by F) has a solution by Lemma 8.2.1.

Theorem 8.2.1 is proved. □

Let us sketch an alternative derivation of the inequality (8.2.12) using the maximum principle. Let us derive (8.2.10). The derivation of (8.2.11) is similar.

Assume that (8.2.10) fails. Then $u > a$ at some point in D. Therefore at a point y, at which u attains its maximum value, one has $u(y) \geq u(x)$ for all $x \in D$ and $u(y) > a$. The function u attains its maximum value, which is positive, at some point in D, because u is continuous, vanishes at the boundary of D, and is positive at some point of D by the assumption $u > a$. At the point y, where the function u attains its maximum, one has $-\Delta u \geq 0$ and $k^2 u(y) > 0$. Moreover, $f(u(y)) > 0$ by the assumption (8.2.1), since $u(y) > a$. Therefore the left-hand side of equation (8.2.2) is positive, while its left-hand side is zero. Thus, we have got a contradiction, and estimate (8.2.10) is proved. Similarly one proves estimate (8.2.11). Thus, (8.2.12) is proved. □

Chapter 9

DSM in Banach spaces

In this chapter we generalize some of our results to operator equations in Banach spaces.

9.1 Well-posed problems

Consider the equation

$$F(u) = f, \tag{9.1.1}$$

where $F : X \to Y$ is a map from a Banach space X into a Banach space Y.

Consider first the case when both assumptions (1.3.1) and (1.3.2) hold:

$$\sup_{u \in B(u_0,R)} ||[F'(u)]^{-1}|| \leq m(R), \tag{9.1.2}$$

$$\sup_{u \in B(u_0,R)} ||F^{(j)}(u)|| \leq M_j(R), \quad 0 \leq j \leq 2. \tag{9.1.3}$$

In Section 9.3 we use assumption (9.1.3) with $j \leq 3$.

We construct the Newton-type DSM:

$$\dot{u} = -[F'(u)]^{-1}[F(u) - f], \quad u(0) = u_0, \tag{9.1.4}$$

which makes sense due to (9.1.2). As in Sections 3.1 - 3.2, the right-hand side of (9.1.4) satisfies a Lipschitz condition due to assumptions (9.1.2)-(9.1.3). Therefore problem (9.1.4) has a unique local solution. As in Section

2.6, Lemma 2.6.1, this local solution is a global one, provided that bound (2.6.8) is established with the constant c independent of time. Denote

$$g(t) :=< F(u(t)) - f, h >,$$

where $< w, h >$ is the value of a linear functional $h \in Y^*$ at the element $F(u(t)) - f$.

We have, using equation (9.1.4),

$$\dot{g} =< F'(u)\dot{u}, h >= -g. \tag{9.1.5}$$

Therefore

$$g(t) = g(0)e^{-t}.$$

Taking the supremum over $h \in Y^*$, $||h|| = 1$, one gets

$$G(t) = G(0)e^{-t}, \quad G(t) := ||F(u(t)) - f||. \tag{9.1.6}$$

Using equation (9.1.4), estimate (9.1.2), and formula (9.1.6), one obtains

$$||\dot{u}|| \le m(R)G(0)e^{-t}. \tag{9.1.7}$$

Therefore

$$||u(t) - u(0)|| \le m(R)G(0). \tag{9.1.8}$$

We assume that

$$m(R)||F(u) - f|| \le R. \tag{9.1.9}$$

This assumption ensure that the trajectory $u(t)$ stays inside the ball $B(u_0, R)$ for all $t \ge 0$, so that the local solution to problem (9.1.4) is the global one.

Moreover, (9.1.7) and (9.1.8) imply the existence of $u(\infty)$ and the estimate:

$$||u(t) - u(\infty)|| \le m(R)||F(u_0) - f||e^{-t}. \tag{9.1.10}$$

This estimate shows exponential rate of convergence of $u(t)$ to $u(\infty)$. Finally, passing to the limit $t \to \infty$ in (9.1.6), one checks that $u(\infty)$ solves equation (9.1.1). Let us summarize the result.

Theorem 9.1.1. *Assume that $F : X \to Y$ is a map from a Banach space X into a Banach space Y, and (9.1.2), (9.1.3) and (9.1.9) hold. Then equation (9.1.1) has a solution, problem (9.1.4) has a unique global solution $u(t)$, there exists $u(\infty)$, $u(\infty)$ solves equation (9.1.1), and estimates (9.1.8) and (9.1.10) hold.*

Remark 9.1.1. Theorem 9.1.1 gives a sufficient condition (condition(9.1.9)) for the existence of a solution to equation (9.1.1).

If one knows a priori that equation (9.1.1) has a solution, then condition (9.1.9) is always satisfied if one chooses the initial approximation u_0 sufficiently close to this solution, because then the quantity $||F(u_0) - f||$ can be made as small as one wishes.

The results in Theorem 9.1.1 are similar to the results obtained in Section 3.2 for the equation (9.1.1) with the operator F which acted in a Hilbert space.

9.2 Ill-posed problems

Consider equation (9.1.1) assuming that condition (9.1.3) holds, but condition (9.1.2) does not hold, and that equation (9.1.1) has a solution y, possibly non-unique.

Equation (9.1.4) cannot be used now because the operator $[F'(u)]^{-1}$ is not defined. Therefore we need some additional assumptions on F to treat ill-posed problems.

Let us denote

$$A := F'(u),$$

and make the spectral assumption from Section 8.1:

$$||A_\varepsilon^{-1}|| \leq \frac{c_0}{\varepsilon}, \quad 0 < \varepsilon < r_0, \tag{9.2.1}$$

where c_0 and r_0 are some positive constants.

Theorem 8.1.1 from Section 8.1 claims that assumptions (9.2.1) and (9.1.3) imply existence of a solution to equation (8.1.2). The proof of Theorem 8.1.1 is valid, after suitable modifications, in the case when $F : X \to X$ is an operator from a Banach space X into X. Let us point out these modifications.

Definition (8.1.5) should be replaced by

$$g(t) :=< F(u(t)) + \varepsilon u(t) - f, h >, \tag{9.2.2}$$

where $h \in X^*$ is arbitrary, $||h|| = 1$. We have

$$G(t) := ||F(u(t)) + \varepsilon u(t) - f|| = \sup_{||h||=1,\ h\in X^*} |g(t)|. \tag{9.2.3}$$

As in Section 9.1, we prove that the problem

$$\dot{u} = -A_\varepsilon^{-1}[F(u) + \varepsilon u(t) - f], \quad u(0) = u_0, \tag{9.2.4}$$

has a unique local solution $u(t)$ and

$$g(t) \leq g(0)e^{-t}.$$

Formula (9.2.3) yields

$$G(t) \leq G(0)e^{-t}, \quad G(0) = ||F(u_0) + \varepsilon u_0 - f||. \tag{9.2.5}$$

Equation (9.2.4) and estimates (9.2.1) and (9.2.5) imply

$$||\dot{u}|| \leq \frac{c_0 G(0)}{\varepsilon} e^{-t}. \tag{9.2.6}$$

For any fixed u_0 the inequality

$$\frac{c_0 G(0)}{\varepsilon} \leq R \tag{9.2.7}$$

holds if $\varepsilon > 0$ is sufficiently small.

If (9.2.7) holds, then

$$||u(t) - u(0)|| \leq \frac{c_0 G(0)}{\varepsilon} \leq R, \tag{9.2.8}$$

so that $u(t) \in B(u_0, R)$ for $t \geq 0$. This implies, as in Section 2.6, see Lemma 2.6.1, that the local solution $u(t)$ to (9.2.4) is the global one. Moreover (9.2.4) implies the existence of $u(\infty)$ and the estimate

$$||u(t) - u(\infty)|| \leq \frac{c_0 G(0)}{\varepsilon} e^{-t} \leq Re^{-t}. \tag{9.2.9}$$

Passing to the limit $t \to \infty$ in (9.2.5) we see that $u(\infty)$ solves the equation

$$F(u(\infty)) + \varepsilon u(\infty) = f. \tag{9.2.10}$$

Let us formulate the result we have just proved.

Theorem 9.2.1. *Assume that $F : X \to X$, that (9.2.1), (9.1.3), and (9.2.7) hold. Then problem (9.2.4) has a unique global solution $u(t)$, there exists $u(\infty)$, $u(\infty)$ solves equation (9.2.10), and estimates (9.2.8), (9.2.9) hold.*

The problem is:

Under what assumptions can we establish that $u(t) := u_\varepsilon(t)$ converges, as $\varepsilon \to 0$, to a solution of the limiting equation?

This question is discussed in the next Section.

9.3 Singular perturbation problem

In Section 9.2 we have proved that equation

$$F(u) + \epsilon u = f, \quad 0 < \epsilon < r_0, \tag{9.3.1}$$

has a solution $u = u_\epsilon$. This does not imply, in general, the existence of the solution to the limiting equation

$$F(y) = f \tag{9.3.2}$$

A simple example was given in Section 8.1 below formula (8.1.12).

Therefore we have to assume that equation (9.3.2) has a solution.

If equation (9.3.2) is solvable, then our aim is to give sufficient conditions for a solution $w = w_\epsilon$ to the equation

$$F(w) + \epsilon(w - z) = f \tag{9.3.3}$$

to converge, as $\epsilon \to 0$, to a solution of the limiting equation (9.3.2).

Let us formulate our result.

Theorem 9.3.1. *Assume that equation (9.3.2) is solvable, conditions (9.2.1) and (9.1.3) with $j \le 3$ hold, and z in (9.3.3) is chosen so that*

$$y - z = \tilde{A}v, \quad ||v|| < \frac{1}{2M_2c_0(1 + c_0)},$$

where c_0 is the constant from (9.2.1), and $\tilde{A} = F'(y)$.

Then

$$\lim_{\epsilon \to 0} ||w_\epsilon - y|| = 0, \tag{9.3.4}$$

where y is a solution to (9.3.2).

Proof of Theorem 9.3.1. Using Taylor's formula, let us write equation (9.3.3) as

$$0 = F(w) - F(y) + \epsilon(w - y) + \epsilon(y - z) = \tilde{A}_\epsilon\psi + \eta + \epsilon\tilde{A}v, \tag{9.3.5}$$

where $\tilde{A}_\epsilon := \tilde{A} + \epsilon I$ and

$$\psi = w - y, \quad ||\eta|| \le \frac{M_2||\psi||^2}{2}, \quad \eta = \eta(w). \tag{9.3.6}$$

Equation (9.3.5) can be written as

$$\psi = T\psi, \tag{9.3.7}$$

where

$$T\psi := -\tilde{A}_\epsilon^{-1}\eta - \epsilon\tilde{A}_\epsilon^{-1}\tilde{A}v. \tag{9.3.8}$$

We *claim*, that the operator T maps a ball

$$B_R := \{\psi : ||\psi|| \leq R\}$$

into itself and is a contraction mapping in this ball. If this claim is established, then the contraction mapping principle yields existence and uniqueness of the solution to (9.3.7) in the ball B_R. We choose $R = R(\epsilon)$ so that

$$\lim_{\epsilon\to 0} R(\epsilon) = 0.$$

Thus

$$||w_\epsilon - y|| \leq R(\epsilon) \to 0 \text{ as } \epsilon \to 0, \tag{9.3.9}$$

and Theorem 9.3.1 is proved.

Let us verify the *claim*. Note that

$$||\tilde{A}_\epsilon^{-1}\tilde{A}|| = ||\tilde{A}_\epsilon^{-1}(\tilde{A} + \epsilon - \epsilon)|| \leq 1 + c_0.$$

Thus we have $TB_R \subseteq B_R$, provided that

$$||T\psi|| \leq \frac{c_0}{\epsilon}\frac{M_2}{2}R^2 + \epsilon(1 + c_0)||v|| \leq R. \tag{9.3.10}$$

This inequality holds if

$$\frac{\epsilon}{c_0 M_2}(1 - \rho) \leq R \leq \frac{\epsilon}{c_0 M_2}(1 + \rho), \tag{9.3.11}$$

where

$$\rho = \sqrt{1 - 2c_0(1 + c_0)M_2||v||}, \qquad ||v|| < \frac{1}{2c_0(1 + c_0)M_2}. \tag{9.3.12}$$

Let us now verify that T is a contraction mapping in B_R, where

$$R = \frac{\epsilon}{c_0 M_2}(1 - \rho). \tag{9.3.13}$$

Let $p, q \in B_R$. Then

$$Tp - Tq = -\tilde{A}_\epsilon^{-1}[\eta(p) - \eta(q)], \tag{9.3.14}$$

where

$$\eta(p) = \int_0^1 (1-s)F''(y+sp)pp\, ds \tag{9.3.15}$$

is the remainder in the Taylor formula

$$F(p) - F(y) = \tilde{A}(p-y) + \eta(p). \tag{9.3.16}$$

If $p, q \in B_R$, then

$$\begin{aligned} ||\eta(p) - \eta(q)|| &= ||\int_0^1 (1-s)[F''(y+sp)pp - F''(y+sq)qq]ds|| \\ &\le \int_0^1 (1-s)\{|||[F''(y+sp) - F''(y+sq)pp]||| \\ &\quad + ||F''(y+sq)(pp-qq)||\}ds \\ &\le \int_0^1 ds(1-s)(M_3 sR^2 + 2M_2R)||p-q||. \end{aligned} \tag{9.3.17}$$

Thus,

$$||\eta(p) - \eta(q)|| \le \left(M_2R + \frac{M_3R^2}{6}\right)||p-q||. \tag{9.3.18}$$

From (9.3.14), (9.3.18) and (9.2.1) we get

$$||Tp - Tq|| \le c_0 \frac{M_2R}{\epsilon}\left(1 + \frac{M_3R}{6M_2}\right)||p-q||. \tag{9.3.19}$$

This and (9.3.13) imply

$$||Tp - Tq|| \le (1 - \rho + O(\epsilon))||p - q||, \quad \epsilon \to 0, \quad p, q \in B_R. \tag{9.3.20}$$

Therefore, for sufficiently small ϵ, the mapping T is a contraction mapping in B_R, where R is given by (9.3.13).

Theorem 9.3.1 is proved. □

Remark 9.3.1. If $\tilde{A} := F(y) \neq 0$, then one can always choose z so that

$$y - z = \tilde{A}v, \quad ||v|| < b, \tag{9.3.21}$$

where $b > 0$ is an arbitrary small fixed number. Indeed, if $\tilde{A} \neq 0$ then there exists an element $p \neq 0$ such that $\tilde{A}p \neq 0$. Choose z such that $y - z = \lambda\tilde{A}p$, where $\lambda = \text{const}$ and $|\lambda|\ ||p|| < b$. Then $v = \lambda p, \quad ||v|| < b$.

Remark 9.3.2. If one can choose $z = 0$ in Theorem 9.3.1, i.e., if $y = \tilde{A}v$ and $||v||$ is sufficiently small, then the solution to equation (9.3.1) tends to y as $\epsilon \to 0$.

In this case the DSM (9.2.4) yields a solution to equation (9.3.2) by the formula:

$$y = \lim_{\epsilon \to 0} \lim_{t \to \infty} u_\epsilon(t), \tag{9.3.22}$$

where $u_\epsilon(t)$ is the solution to problem (9.2.4).

Chapter 10

DSM and Newton-type methods without inversion of the derivative

In this Chapter we construct the DSM so that there is no need to invert $F'(u)$.

10.1 Well-posed problems

A Newton-type DSM requires calculation of $[F'(u)]^{-1}$, see (3.2.2).

This is a difficult and time-consuming step. Can one avoid such a step? In this Section we assume that conditions (9.1.2) and (9.1.3) hold, and equation

$$F(u) = f \tag{10.1.1}$$

has a solution y.

Consider the following DSM:

$$\dot{u} = -Q[F(u) - f], \quad t \geq 0, \tag{10.1.2}$$

$$\dot{Q} = -TQ + A^*, \tag{10.1.3}$$

$$u(0) = u_0, \quad Q(0) = Q_0, \tag{10.1.4}$$

where

$$A := F'(u), \quad T := A^*A, \tag{10.1.5}$$

$u_0 \in H$, and $Q(t)$ is an operator-valued function.

Our result is the following theorem.

Theorem 10.1.1. *Assume (9.1.2), (9.1.3). Suppose that equation (10.1.1) has a solution y, u_0 is sufficiently close to y, and Q_0 is sufficiently close to $\tilde{A}^{-1}$, where $\tilde{A} := F'(y)$. Then problem (10.1.2) - (10.1.4) has a unique global solution, there exists $u(\infty)$, $u(\infty) = y$, i.e.,*

$$\lim_{t\to\infty} ||u(t) - y|| = 0, \tag{10.1.6}$$

and

$$\lim_{t\to\infty} ||Q(t) - \tilde{A}^{-1}|| = 0. \tag{10.1.7}$$

Proof of Theorem 10.1.1. From our assumptions the existence and uniqueness of the local solution to problem (10.1.2) - (10.1.4) follows. If we derive a uniform with respect to t bound on the norm $||u(t)|| + ||Q(t)|| \le c$, then, as in Section 2.6, we prove that the local solution to (10.1.2) - (10.1.4) is a global one. First, let us estimate the norm $||Q(t)||$ uniformly with respect to t. We use Theorem 5.2.1 and estimate (5.2.4). Since A is invertible, there is a constant $\epsilon = \text{const} > 0$ such that

$$(Th, h) \ge \epsilon ||h||^2. \tag{10.1.8}$$

Therefore estimate (5.2.4) yields

$$||Q(t)|| \le e^{-\epsilon t}[||Q_0|| + \int_0^t M_1 e^{\epsilon s} ds] \le ||Q_0|| + \frac{M_1}{\epsilon} := c_1. \tag{10.1.9}$$

Thus, $||Q(t)||$ is bounded uniformly with respect to t.

Let us now estimate $||u(t)||$. We denote

$$w := u(t) - y, \quad ||w(t)|| := g(t),$$

and use Taylor's formula:

$$F(u) - f = F(u) - F(y) = \tilde{A}w + K(w), \tag{10.1.10}$$

where

$$||K(w)|| \le \frac{M_2}{2}||w||^2,$$

and M_2 is the constant from (9.1.3). Thus equation (10.1.2) can be written as

$$\dot{w} = -Q\tilde{A}w - QK(w). \tag{10.1.11}$$

Define

$$\Lambda := I - Q\tilde{A}. \tag{10.1.12}$$

Then

$$\dot{w} = -w + \Lambda w - QK(w). \tag{10.1.13}$$

Multiply this equation by w and get

$$g\dot{g} \leq -g^2 + \lambda g^2 + \frac{c_1 M_2}{2} g^3,$$

and

$$\dot{g} \leq -\gamma g + c_2 g^2, \quad \gamma := -1 - \lambda, \quad \gamma \in (0,1). \tag{10.1.14}$$

Here we have used the estimate

$$(\Lambda w, w) \leq \lambda ||w||^2, \quad \lambda \in (0,1), \tag{10.1.15}$$

which will be proved later.

From (10.1.14) one derives

$$g(t) \leq r e^{-\gamma t}, \quad r := \frac{g(0)}{1 - g(0)c_2}, \tag{10.1.16}$$

and we assume that

$$g(0)c_2 < 1. \tag{10.1.17}$$

Assumption (10.1.17) is always justified if $g(0)$ is sufficiently small, i.e. if $||u_0 - y||$ is sufficiently small.

Let us verify inequality (10.1.15). Using definition (10.1.12) and differentiating it with respect to time, one gets:

$$\dot{\Lambda} = -\dot{Q}\tilde{A}. \tag{10.1.18}$$

This equation, (10.1.12) and (10.1.3) yield:

$$\dot{\Lambda} = (TQ - A^*)\tilde{A} = T(I - \Lambda) - A^*\tilde{A} = -T\Lambda + A^*(A - \tilde{A}). \tag{10.1.19}$$

We have

$$||A^*(A - \tilde{A})|| \leq M_1 M_2 g(t) \leq M_1 M_2 r e^{-\gamma t}, \tag{10.1.20}$$

where r is given by (10.1.16).

Let us apply estimate (5.2.4) and inequality (10.1.8) to equation (10.1.19). This yields:

$$||\Lambda(t)|| \le e^{-\epsilon t}\left(||\Lambda(0)|| + \int_0^t M_1 M_2 r e^{-\gamma s} e^{\epsilon s} ds\right). \tag{10.1.21}$$

Thus

$$||\Lambda(t)|| \le ||\Lambda(0)|| + cg(0) := \lambda, \tag{10.1.22}$$

where

$$c := \frac{M_1 M_2}{1 - g(0)c_2} \sup_{t \ge 0} \frac{e^{-\gamma t} - e^{-\epsilon t}}{\epsilon - \gamma}. \tag{10.1.23}$$

Estimate (10.1.22) shows that $\lambda \in (0,1)$ if $||\Lambda(0)||$ and $||u_0 - y||$ are sufficiently small.

To complete the proof of Theorem 10.1.1 we need to verify formula (10.1.17). Estimate (10.1.21) implies:

$$\lim_{t\to\infty} ||\Lambda(t)|| \le O(e^{-\min(\epsilon,\gamma)t}) \to 0 \text{ as } t \to \infty. \tag{10.1.24}$$

This formula and (10.1.12) imply

$$\lim_{t\to\infty} ||Q(t)\tilde{A} - I|| = 0,$$

and since $\tilde{A}^{-1}$ is a bounded operator, we get

$$\lim_{t\to\infty} ||Q(t) - \tilde{A}^{-1}|| = 0. \tag{10.1.25}$$

Theorem 10.1.1 is proved. □

10.2 Ill-posed problems

In this Section we assume (9.1.3), but not (9.1.2). Let

$$F'(u) := A, \quad T := A^*A, \quad \tilde{T} = \tilde{A}^*\tilde{A}, \quad \tilde{A} := F'(y), \tag{10.2.1}$$

where y is a solution to the equation

$$F(u) = f. \tag{10.2.2}$$

Consider the following DSM:

$$\dot{u} = -Q[A^*(F(u) - f) + \epsilon(t)(u - z)], \tag{10.2.3}$$

$$\dot{Q} = -T_{\epsilon(t)}Q + I, \tag{10.2.4}$$

$$u(0) = u_0, \quad Q(0) = Q_0, \tag{10.2.5}$$

where z is an element which we choose later.

Assume that:

$$0 < \epsilon(t) \searrow 0, \quad \frac{|\dot{\epsilon}|}{\epsilon} \leq \frac{\gamma}{2}, \quad \frac{|\dot{\epsilon}(t)|}{\epsilon^2(t)} \leq b, \quad \lim_{t\to\infty} \frac{|\dot{\epsilon}(t)|}{\epsilon^2(t)} = 0,$$

$$\int_0^\infty \epsilon(s)ds = \infty, \tag{10.2.6}$$

where γ is defined in formula (10.2.14) below, and

$$||w(t)|| = g(t), \quad b = \text{const} > 0.$$

Let

$$w := u(t) - y,$$

and assume that z is chosen so that

$$y - z = \tilde{T}v, \quad ||v|| \leq \frac{\lambda\gamma}{4c_2\epsilon^2(0)}, \tag{10.2.7}$$

see inequality (10.2.21) below.

Define

$$\Lambda := I - Q\tilde{T}_{\epsilon(t)}. \tag{10.2.8}$$

From equation (10.2.4) and estimate (5.2.4) we get:

$$\begin{aligned} ||Q(t)|| &\leq e^{-\int_0^t \epsilon(s)ds}\left[||Q(0)|| + \int_0^t e^{-\int_0^s \epsilon(p)dp}ds\right] \\ &\leq ||Q(0)|| + \frac{1}{\epsilon(t)} \leq \frac{c_1}{\epsilon(t)} \end{aligned} \tag{10.2.9}$$

Let us write equation (10.2.3) as

$$\dot{w} = -w + \Lambda w + Q(\tilde{A}^* - A^*)\tilde{A}w + Q(\tilde{A}^* - A^*)K - \epsilon Q\tilde{T}v, \tag{10.2.10}$$

where v is defined (10.2.7) and K is defined in (10.1.10). We have

$$||\tilde{A}|| \leq M_1, \quad ||\tilde{A}^* - A^*|| \leq M_2 g, \quad ||Q|| \leq \frac{c_1}{\epsilon(t)}, \quad ||w|| := g,$$

so

$$||Q(\tilde{A}^* - A^*)\tilde{A}w|| \le \frac{c_1}{\epsilon(t)} M_1 M_2 g^2(t),$$

and

$$||Q(\tilde{A}^* - A^*)K|| \le \frac{c_1}{\epsilon(t)} 2M_1 \frac{M_2}{2} g^2(t) = \frac{c_1 M_1 M_2 g^2}{\epsilon(t)}. \tag{10.2.11}$$

We will prove below that

$$||Q\tilde{T}|| \le c_2, \tag{10.2.12}$$

and

$$||\Lambda|| \le \lambda_0 < 1, \tag{10.2.13}$$

where λ_0 is a positive constant independent of t.

We define constant γ by the formula:

$$\gamma := 1 - \lambda_0, \quad \gamma \in (0,1). \tag{10.2.14}$$

Multiply (10.2.10) by w and use estimates (10.2.11) - (10.2.14) to get

$$\dot{g}g \le -\gamma g^2 + \frac{c_0}{\epsilon(t)} g^3 + c_2 \epsilon(t) ||v|| g, \tag{10.2.15}$$

where

$$c_0 := 2c_1 M_1 M_2, \tag{10.2.16}$$

and c_2 is the constant from (10.2.12).

Since $g(t) \ge 0$, we obtain from (10.2.15) the inequality

$$\dot{g} \le -\gamma g(t) + \frac{c_0}{\epsilon(t)} g^2 + c_2 \epsilon(t) ||v||, \quad \gamma \in (0,1). \tag{10.2.17}$$

We apply to this differential inequality Theorem 5.1.1. Let us check conditions (5.1.2) - (5.1.4), which are the assumptions in this theorem. We take

$$\mu = \frac{\lambda}{\epsilon(t)}, \quad \text{where} \quad \lambda = \text{const} > 0.$$

Condition (5.1.2) for inequality (10.2.17) takes the form:

$$\frac{c_0}{\epsilon(t)} \le \frac{\lambda}{2\epsilon(t)} \left(\gamma - \frac{|\dot{\epsilon}(t)|}{\epsilon(t)} \right). \tag{10.2.18}$$

Using (10.2.6) one concludes that this inequality is satisfied if

$$\frac{4c_0}{\gamma} \leq \lambda. \tag{10.2.19}$$

Condition (5.1.3) for inequality (10.2.17) is:

$$c_0\epsilon(t)||v|| \leq \frac{\lambda}{2\epsilon(t)}\left(\gamma - \frac{|\dot{\epsilon}(t)|}{\epsilon(t)}\right). \tag{10.2.20}$$

This inequality is satisfied if

$$\frac{4c_2}{\lambda\gamma}\epsilon^2(0)||v|| \leq 1. \tag{10.2.21}$$

Inequality (10.2.21) is satisfied if $||v||$ is sufficiently small. Finally, condition (5.1.4) is

$$||u_0 - y||\frac{\lambda}{\epsilon(0)} < 1. \tag{10.2.22}$$

Inequality (10.2.22) holds if $||u_0 - y||$ is sufficiently small.

Thus, if

$$\lambda \geq \frac{4c_0}{\gamma},$$

then condition (10.2.19) holds; if

$$||v|| \leq \frac{\lambda\gamma}{4c_2\epsilon^2(0)},$$

then condition (10.2.21) holds; and, finally, if

$$||u_0 - y|| < \frac{\epsilon(0)}{\lambda},$$

then condition (10.2.22) holds. If these conditions hold, then the inequality (5.1.6) yields:

$$||u(t) - y|| < \frac{\epsilon(t)}{\lambda}. \tag{10.2.23}$$

This estimate yields a uniform bound on the norm $||u(t)||$, and therefore implies the global existence of the solution to problem (10.2.3) - (10.2.5), the existence of $u(\infty)$ and the relation $u(\infty) = y$, where y solves equation (10.2.2).

To complete the argument we have to prove estimates (10.2.12) and (10.2.13). Estimate (10.2.12) follows from (10.2.13) and from the definition (10.2.8). Let us prove (10.2.13). Using definition (10.2.8) we derive:

$$\dot{\Lambda} = -\dot{Q}\tilde{T}_{\epsilon(t)} - Q\dot{\epsilon}. \tag{10.2.24}$$

Using equations (10.2.4) and (10.2.24), we obtain

$$\dot{\Lambda} = -T_{\epsilon(t)}\Lambda + T_{\epsilon(t)} - \tilde{T}_{\epsilon(t)} - Q\dot{\epsilon}. \tag{10.2.25}$$

This equation and Theorem 5.2.1 yield:

$$\begin{aligned} ||\Lambda(t)|| \quad \leq \quad & e^{-\int_0^t \epsilon(s)ds}\Big[||\Lambda(0) + \int_0^t e^{-\int_0^s \epsilon(p)dp}(||T_{\epsilon(t)} - \tilde{T}_{\epsilon(t)}|| \\ & + ||Q(s)||\ |\dot{\epsilon}(s)|)ds\Big] \end{aligned} \tag{10.2.26}$$

Using inequality (10.2.23), we derive the following estimate:

$$\begin{aligned} ||T_{\epsilon(t)} - \tilde{T}_{\epsilon(t)}|| \quad & = \quad ||A^*A - \tilde{A}^*\tilde{A}|| \leq 2M_1M_2||u(t) - y|| \\ & \leq \quad \frac{2M_1M_2\epsilon(t)}{\lambda}. \end{aligned} \tag{10.2.27}$$

Using (10.2.9), we get

$$||Q(s)||\ |\dot{\epsilon}(s)| \leq c_1\frac{|\dot{\epsilon}(s)|}{\epsilon(s)}. \tag{10.2.28}$$

From (10.2.26) - (10.2.28) we obtain:

$$||\Lambda(t)|| \leq ||\Lambda(0)|| + \frac{2M_1M_2}{\lambda} + b, \tag{10.2.29}$$

where the following estimate

$$e^{-\int_0^t \epsilon(s)ds}\int_0^t \epsilon(s)e^{-\int_0^s \epsilon(p)dp}ds = 1 - e^{-\int_0^t \epsilon(s)ds} \leq 1 \tag{10.2.30}$$

was used.

Assume that

$$||\Lambda(0)|| + \frac{2M_1M_2}{\lambda} + b \leq \lambda_0 < 1. \tag{10.2.31}$$

Then condition (10.2.13) holds and, therefore, inequality (10.2.12) holds. Condition (10.2.31) holds if

$$||I - Q(0)\tilde{T}_{\epsilon(0)}|| < 1.$$

Indeed, one can choose $\epsilon(t)$ such that $b > 0$ is as small as one wishes. For example, if

$$\epsilon(t) = \frac{C}{(a+t)^q},$$

then

$$\frac{|\dot{\epsilon}|}{\epsilon^2} = \frac{q}{C(a+t)^{1-q}} \leq \frac{q}{Ca^{1-q}}, \quad 0 < q < 1, \tag{10.2.32}$$

and this number can be made as small as one wishes by taking C sufficiently large. The term $\frac{2M_1M_2}{\lambda}$ can be made as small as one wishes by choosing λ sufficiently large.

Let us formulate the result we have proved.

Theorem 10.2.1. *Assume that equation (10.2.2) has a solution y, possibly non-unique, that $\epsilon(t)$ satisfies (10.2.6), and that $||u_0 - y||$ and $||I - Q(0)\tilde{T}_{\epsilon(0)}||$ are sufficiently small. Then problem (10.2.3) - (10.2.5) has a unique global solution and estimate (10.2.23) holds.*

Remark 10.2.1. Estimate (10.2.23) implies that $u(\infty) = y$ solves equation (10.2.2).

Remark 10.2.2. In contrast to the well-posed case (see Theorem 10.1.1 from Section 10.1), we do not prove in the case of ill-posed problems the convergence of $Q(t)$ as $t \to \infty$. The reason is: the inverse of the limiting operator T^{-1} does not exist, in general.

Remark 10.2.3. If $||\tilde{T}|| < \infty$, then the condition

$$||I - Q(0)\tilde{T}_{\epsilon(0)}|| < 1$$

can be satisfied algorithmically in spite of the fact that we do not know y and, consequently, we do not know $\tilde{T}_{\epsilon(0)}$. For example, if one takes

$$Q(0) = cI,$$

where $c > 0$ is constant, and denote $\epsilon(0) := \nu$, then

$$||I - Q(0)\tilde{T}_{\epsilon(0)}|| = \sup_{0 \leq s \leq ||\tilde{T}||} \left| 1 - \frac{c}{\nu + s} \right| = 1 - \frac{c}{\nu + ||\tilde{T}_{\epsilon(0)}||} < 1,$$

provided that

$$\frac{c}{\nu + ||\tilde{T}_{\epsilon(0)}||} < 1.$$

Chapter 11

DSM and unbounded operators

In this Chapter we consider the case when the operator F in the equation $F(u) = 0$ is unbounded.

11.1 Statement of the problem

Consider the equation

$$G(u) := Lu + g(u) = 0, \qquad (11.1.1)$$

where L is a linear closed, unbounded, densely defined operator in a Hilbert space H, which has a bounded inverse,

$$||L^{-1}|| \leq m, \qquad (11.1.2)$$

while g is a nonlinear operator which satisfies assumptions (1.3.2). The operator G in (11.1.1) is not continuously Fréchet differentiable in the standard sense if L is unbounded. Recall that the standard definition of the Fréchet derivative at a point u requires the existence of a bounded linear operator $A(u)$ such that

$$F(u+h) = F(u) + A(u)h + o(||h||) \quad \text{as } ||h|| \to 0. \qquad (11.1.3)$$

If G is defined by formula (11.1.1), then formally

$$G'(u) := A := A(u) = L + g'(u),$$

but in the standard definition the element $h \in H$ in (11.1.3) is arbitrary, subject to the condition $||h|| \to 0$. In the case of unbounded G, defined in (11.1.1), the element h cannot be arbitrary: it has to belong to the domain

$D(L)$ of L. The operator $A = L + g'(u)$ has the domain $D(A) = D(L)$ dense in H, $D(L)$ is a linear manifold in H. If assumption (11.1.2) holds, then equation (11.1.1) is equivalent to

$$F(u) := u + L^{-1}g(u) = 0. \tag{11.1.4}$$

To this equation we can apply the theory developed for equations which satisfy assumption (1.3.2). We have

$$F'(u) = I + L^{-1}g(u), \qquad \sup_{u \in B(u_0,R)} ||F'(u)|| \le M_1(R), \tag{11.1.5}$$

and

$$\sup_{u \in B(u_0,R)} ||F''(u)|| \le M_2(R).$$

If equation (11.1.1) has a solution, then this solution solves equation (11.1.4), so we may apply the results of Chapters 3, 6 - 10 to equation (11.1.4). For example, Theorem 3.2.1 from Section 3.2 yields the following result.

Theorem 11.1.1. *Assume that equation (11.1.1) is solvable, L is a densely defined, closed linear operator, (11.1.2) holds, and the operator F, defined in (11.1.4), satisfies assumptions (1.3.1), (1.3.2), and (3.2.6).*

Then the problem

$$\dot{u} = -[F'(u)]^{-1}F(u), \quad u(0) = u_0, \tag{11.1.6}$$

has a unique global solution $u(t)$, there exists $u(\infty)$, and $F(u(\infty)) = 0$.

If condition (1.3.1) is not satisfied, then problem (11.1.4) can be treated by the methods developed in Chapters 6 - 10.

An example of such treatment is given in Section 11.2.

Example 11.1.1. Consider a semilinear elliptic problem:

$$-\Delta u + g(u) = f \text{ in } D, \quad u|_S = 0, \tag{11.1.7}$$

where $D \subset \mathbb{R}^3$ is a bounded domain with a smooth boundary S, $f \in L^2(D)$ is a given function, $g(u)$ is a smooth nonlinear function and we assume that

$$g(u) \ge 0, \quad g'(u) \ge 0. \tag{11.1.8}$$

Then one can check easily that problem (11.1.7) has no more than one solution. If the solution u of problem (11.1.7) exists, then it solves the problem

$$F(u) := u + L^{-1}g(u) = h, \quad h := L^{-1}f, \tag{11.1.9}$$

where $L^{-1} := (-\Delta)^{-1}$, and $-\Delta$ is the Dirichlet Laplacian. It is well known that estimate (11.1.2) holds for $L^{-1} = (-\Delta)^{-1}$ in a bounded domain. The operator F in (11.1.9) satisfies conditions (1.3.1) and (1.3.2).

Thus, one can solve equation (11.1.9) by the DSM (11.1.6).

11.2 Ill-posed problems

In this Section we consider equation (11.1.1) under the assumption (11.1.2), and the equivalent equation (11.1.4), and we assume that the operator F satisfies condition (1.3.2) but not (1.3.1). We assume that equations (11.1.1) and, therefore, equation (1.1.4) have a solution y, and

$$\tilde{A} := F'(y) \neq 0.$$

Under these assumptions we may apply Theorem 7.1.1.

Consider the following equation

$$F(u) := u + B(u) = f, \quad B(u) := L^{-1}g(u), \tag{11.2.1}$$

which is equation (11.1.4), and the DSM for solving this equation:

$$\dot{u} = -T^{-1}_{a(t)}\left[A^*(F(u) - f) + a(t)(u - z)\right], \quad u(0) = u_0, \tag{11.2.2}$$

where the notations are the same as in Section 7.1, see (7.1.4).

Theorem 7.1.1 from Section 7.1 gives sufficient conditions for the existence and uniqueness of the global solution $u(t)$ to problem (11.2.2), for the existence of $u(\infty) = y$, where $F(y) = 0$.

Theorem 7.2.1 from Section 7.2, applied to equation (11.2.1) with noisy data f_δ, $||f_\delta - f|| \leq \delta$, given in place of the exact data f, allows us to use DSM (11.2.2) with f_δ in place of f. In particular, this Theorem yields the existence of the stopping time t_δ such that the element $u_\delta(t_\delta)$ converges to y as $\delta \to 0$. Here $u_\delta(t)$ is the solution to problem (11.2.2) with f_δ in place of f and y is a solution to equation (11.2.1).

Chapter 12

DSM and nonsmooth operators

In this Chapter we study the equation $F(u) = 0$ with monotone operator F without assuming that $F \in C^2_{\text{loc}}$.

12.1 Formulation of the results

Consider the equation

$$F(u) = 0, \tag{12.1.1}$$

where *F is a monotone operator in a Hilbert space H*:

$$(F(u) - F(v), u - v) \geq 0, \quad \forall u, v \in H. \tag{12.1.2}$$

We do not assume in this Chapter that $F \in C^2_{\text{loc}}$. We assume that *F is defined on all of H and is hemicontinuous.*

Definition 12.1.1. *F is called hemicontinuous if $s_n \to +0$ implies*

$$F(u + s_n v) \rightharpoonup F(u) \quad \forall u, v \in H,$$

and demicontinuous if $u_n \to u$ implies $F(u_n) \rightharpoonup F(u)$. Here $\rightharpoonup$ denotes weak convergence in H.

We also assume that equation (12.1.1) has a solution, and denote by y the unique minimal-norm solution to equation (12.1.1). This solution is well-defined due to Lemmas 6.1.2 and 6.1.3 from Section 6.1. In Lemma 6.1.2 we assumed that F was continuous, but the conclusion of this lemma remain valid if F is hemicontinuous and monotone.

Lemma 12.1.1. *If F is hemicontinuous and monotone, then the set*

$$\mathcal{N}_f := \{u : F(u) = f\}$$

is closed and convex.

Proof. Let $u_n \to u$ and $F(u_n) = f$. We want to prove that $F(u) = f$, i.e., the set $\mathcal{N}_f$ is closed. Since F is monotone we have:

$$(F(u_n) - F(u - tz), u_n - u + tz) \geq 0, \quad t = \text{const} > 0, \quad \forall z \in H. \tag{12.1.3}$$

Taking $n \to \infty$ in (12.1.3) we get:

$$(f - F(u - tz), tz) \geq 0. \tag{12.1.4}$$

Let us take $t \to +0$ and use the hemicontinuity of F. The result is:

$$(f - F(u), z) \leq 0. \tag{12.1.5}$$

Since z is arbitrary, it follows that $F(u) = f$. To check that $\mathcal{N}_f$ is convex we use the argument used in the proof of Lemma 6.1.2.

Lemma 12.1.1 is proved. □

Lemma 12.1.2. *If F is hemicontinuous and monotone, then F is demicontinuous.*

Proof. Let $u_n \to u$. We want to prove that $F(u_n) \rightharpoonup F(u)$. Since F is monotone and $D(F) = H$, it is locally bounded in H (see e.g. [De], p.97). Thus,

$$||F(u_n)|| \leq c,$$

where c does not depend on n. Bounded sets in H are weakly precompact. Consequently, we may assume that

$$F(u_n) \rightharpoonup f.$$

Let us prove that $f = F(u)$. This will complete the proof of Lemma 12.1.2. By the monotonicity of F, we have inequality (12.1.3). Let $n \to \infty$ in (12.1.3) and get

$$(f - F(u - tz), z) \geq 0, \quad \forall z \in H. \tag{12.1.6}$$

Taking $t \to +0$ and using hemicontinuity of F, we obtain

$$(f - F(u), z) \geq 0. \tag{12.1.7}$$

Since z is arbitrary, inequality (12.1.7) implies

$$F(u) = f.$$

Lemma 12.1.2 is proved. □

Consider the following DSM for solving equation (12.1.1):

$$\dot{u} = -F(u) - au, \quad u(0) = u_0, \quad a = \text{const} > 0. \tag{12.1.8}$$

Lemma 12.1.3. *Assume that F is monotone, hemicontinuous, and $D(F) = H$. Then problem (12.1.8) has a unique global solution for any $u_0 \in H$ and any $a \geq 0$.*

Proof. In this proof we use the known argument based on Peano approximations (see e.g [De], p.99, and [R38]). Uniqueness of the solution to (12.1.8) can be proved easily: if $u(t)$ and $v(t)$ solve (12.1.8) then $w := u - v$ solves the problem

$$\dot{w} = -[F(u) - F(v)] - aw, \quad w(0) = 0. \tag{12.1.9}$$

Multiply (12.1.9) by w and use the monotonicity of F to get

$$(\dot{w}, w) \leq -a(w, w), \quad w(0) = 0, \quad a = \text{const} > 0. \tag{12.1.10}$$

Integrating (12.1.10) yields $w(t) = 0 \quad \forall t > 0$, i.e. $u = v$. To prove the existence of the solution to (12.1.8), we define the solution to (12.1.8) as the solution to the equation

$$u(t) = u_0 - \int_0^t [F(u(s)) + au(s)]ds. \tag{12.1.11}$$

We prove the existence of the solution to (12.1.11) with $a = 0$. The proof for $a > 0$ is even simpler. Consider the solution to the following equation

$$u_n(t) = u_0 - \int_0^t F(u_n(s - \frac{1}{n}))ds, \quad u_n(t) = u_0 \text{ for } t \leq 0. \tag{12.1.12}$$

From (12.1.12) it follows that

$$||u_n(t) - u_0|| \leq ct, \quad 0 \leq t \leq \frac{r}{c}, \tag{12.1.13}$$

where

$$c = \sup_{u \in B(u_0, r)} ||F(u)||, \quad B(u_0, r) := \{u : ||u - u_0|| \leq r\},$$

and $c < \infty$ because of the local boundedness of monotone operators defined on all of H. From equation (12.1.8) (with $a = 0$) it follows that $||\dot{u}|| \leq c$. Define

$$z_{nm} := u_n(t) - u_m(t), \quad ||z_{nm}|| = g_{nm}(t). \tag{12.1.14}$$

Equation (12.1.8) with $a = 0$ implies:

$$g_{nm}(t)\dot{g}_{nm}(t) = \\ -(F\left(u_n\left(t - \tfrac{1}{n}\right)\right) - F\left(u_m\left(t - \tfrac{1}{m}\right)\right), u_n(t) - u_m(t)) := J. \quad (12.1.15)$$

One has

$$\begin{aligned} J = &-(F\left(u_n\left(t - \tfrac{1}{n}\right)\right) - F\left(u_m\left(t - \tfrac{1}{m}\right)\right), u_n\left(t - \tfrac{1}{n}\right) - u_m\left(t - \tfrac{1}{m}\right)) \\ &-(F\left(u_n\left(t - \tfrac{1}{n}\right)\right) - F\left(u_m\left(t - \tfrac{1}{m}\right)\right), u_n(t) - u_n\left(t - \tfrac{1}{n}\right) - \\ &\left(u_m(t) - u_m\left(t - \tfrac{1}{m}\right)\right)). \end{aligned}$$

Using the monotonicity of F and the estimates

$$||F(u)|| \le c, \quad \text{and} \quad ||\dot{u}|| \le c,$$

we obtain:

$$\begin{aligned} J &\le \left(\left\|F\left(u_n\left(t - \frac{1}{n}\right)\right)\right\| + \left\|F\left(u_m\left(t - \frac{1}{m}\right)\right)\right\|\right) c \left(\frac{1}{n} + \frac{1}{m}\right) \\ &\le 2c^2\left(\frac{1}{n} + \frac{1}{m}\right). \end{aligned}$$

Thus (12.1.15) implies

$$\frac{d}{dt} g_{nm}^2(t) \le 4c^2\left(\frac{1}{n} + \frac{1}{m}\right) \to 0 \quad \text{as } n, m \to \infty; \quad g_{nm}(0) = 0. \quad (12.1.16)$$

Consequently we obtain

$$\lim_{n,m\to\infty} g_{nm}(t) = 0, \qquad 0 \le t \le \frac{r}{c}. \qquad (12.1.17)$$

From (12.1.17) we conclude that the following limit exists

$$\lim_{n\to\infty} u_n(t) = u(t), \qquad 0 \le t \le \frac{r}{c}. \qquad (12.1.18)$$

Since F is demicontinuous, this implies

$$F(u_n(t)) \rightharpoonup F(u(t)). \qquad (12.1.19)$$

Using (12.1.18) and (12.1.19), let us pass to the limit $n \to \infty$ in (12.1.12) and obtain

$$u(t) = u_0 - \int_0^t F(u(s))ds, \qquad t \le \frac{r}{c}. \qquad (12.1.20)$$

Here we have used the relation

$$F\left(u_n\left(t-\frac{1}{n}\right)\right) \rightharpoonup F(u(t)) \quad \text{as } n \to \infty. \tag{12.1.21}$$

This relation follows from the formula

$$u_n\left(t-\frac{1}{n}\right) \to u(t) \quad \text{as } n \to \infty. \tag{12.1.22}$$

Indeed

$$||u_n\left(t-\frac{1}{n}\right) - u(t)|| \le ||u_n\left(t-\frac{1}{n}\right) - u_n(t)|| + ||u_n(t) - u(t)||.$$

The second term on the right converges to zero due to (12.1.18), and the first term is estimated as follows:

$$||u_n\left(t-\frac{1}{n}\right) - u_n(t)|| \le c\frac{1}{n} \to 0 \quad \text{as } n \to \infty, \tag{12.1.23}$$

where c is the constant from (12.1.13). We can differentiate the solution $u(t)$ of the equation (12.1.20) in the weak sense because convergence in (12.1.19) is the weak convergence. Therefore (12.1.20) implies that

$$\dot{u} = -F(u), \quad u(0) = u_0, \tag{12.1.24}$$

where the derivative $\dot{u}$ is understood in the weak sense, i.e.,

$$\frac{d}{dt}(u(t), \eta) = -(F(u(t)), \eta) \quad \forall \eta \in H.$$

If we would assume that F is continuous rather than hemicontinuous, then the derivative $\dot{u}$ could be understood in the strong sense.

Let us prove that the solution $u(t)$ of the equation (12.1.20) exists for all $t \ge 0$. Assume the contrary: the solution $u(t)$ to (12.1.24) exists on the interval $[0, T]$ but does not exist on $[0, T+d]$, where $d > 0$ is arbitrarily small. This assumption leads to a contradiction if we prove that the limit

$$\lim_{t \to T-0} u(t) = u(T)$$

exists and is finite, because in this case one can take $u(T)$ as the initial data at $t = T$ and construct the solution to equation (12.1.24) on the interval $[T, T+d]$, $d > 0$, so that the solution $u(t)$ will exist on $[0, T+d]$, contrary to our assumption.

To prove that the finite limit $u(T)$ exists, let

$$u(t+h) - u(t) := z(t), \quad ||z(t)|| = g(t), \quad t+h < T. \tag{12.1.25}$$

We have

$$\dot{z} = -[F(u(t+h)) - F(u(t))], \quad z(0) = u(h) - u(0). \tag{12.1.26}$$

Multiply (12.1.26) by z and use the monotonicity of F, to get

$$(\dot{z}, z) \leq 0, \quad z(0) = u(h) - u(0). \tag{12.1.27}$$

Thus,

$$||u(t+h) - u(t)|| \leq ||u(h) - u(0)||. \tag{12.1.28}$$

We have

$$\lim_{h \to 0} ||u(h) - u(0)|| = 0.$$

Therefore,

$$\lim_{h \to 0} ||u(t+h) - u(t)|| = 0. \tag{12.1.29}$$

This relation holds uniformly with respect to t and $t+h$ such that $t+h < T$ and $t < T$.

By the Cauchy criterion for convergence, relation (12.1.29) implies the existence of a finite limit

$$\lim_{t \to T-0} u(t) := u(T). \tag{12.1.30}$$

Lemma 12.1.3 is proved. □

Lemma 12.1.4. *If we denote by $u_a(t)$ the unique solution to (12.1.8), then*

$$\lim_{a \to 0} ||u_a(t) - u(t)|| = 0 \tag{12.1.31}$$

uniformly with respect to $t \in [0, T]$. Here $T > 0$ is an arbitrary large fixed number, and $u(t)$ solves (12.1.24).

Proof. First, we check that

$$\sup_{t \geq 0} ||u_a(t)| \leq c, \tag{12.1.32}$$

where the constant c does not depend on t. To prove (12.1.32), we start with the equation

$$\dot{u}_a = -[F(u_a) - F(0)] - F(0) - au_a,$$

multiply it by u_a and denote

$$g(t) := ||u_a(t)||.$$

Then, using the monotonicity of F, we get

$$g\dot{g} \leq c_0 g - ag^2, \quad c_0 = ||F(0)||.$$

Thus

$$||u_a(t)|| = g(t) \leq g(0)e^{-at} + \frac{c_0(1-e^{-at})}{a} \leq g(0) + \frac{c_0}{a}. \tag{12.1.33}$$

To prove (12.1.31), denote

$$w(t) := u_a(t) - u(t).$$

Then

$$\dot{w} = -[F(u_a) - F(u)] - au_a, \quad w(0) = 0. \tag{12.1.34}$$

Multiply this equation by w, and use the monotonicity of F, to get

$$\begin{aligned} p\dot{p} &\leq -a(u_a, w) \leq a||u_a(t)||p(t), \\ p &:= ||w(t)|| = ||u_a(t) - u(t)||, \quad p(0) = 0. \end{aligned} \tag{12.1.35}$$

Thus

$$p(t) \leq a \int_0^t ||u_a(s)||ds. \tag{12.1.36}$$

If $t \in [0, T]$, then the first inequality (12.1.33) implies

$$\max_{0\leq t\leq T} a||u_a(t)|| \leq ag(0) + c_0 \max_{0\leq t\leq T} |1 - e^{-at}|. \tag{12.1.37}$$

Passing to the limit $a \to 0$ in (12.1.37) we obtain

$$\lim_{a\to 0} \max_{0\leq t\leq T} a||u_a(t)|| = 0. \tag{12.1.38}$$

From (12.1.38) and (12.1.36) the relation (12.1.31) follows.

Lemma 12.1.4 is proved. □

Let us now state our main results.

Theorem 12.1.1. *Assume that equation $F(u) = 0$ has a solution, possibly non-unique, that $F : H \to H$ is monotone, continuous, $D(F) = H$, and $a = const > 0$. Then problem (12.1.8) has a unique global solution $u_a(t)$ and*

$$\lim_{a\to 0}\lim_{t\to\infty} ||u_a(t) - y|| = 0, \tag{12.1.39}$$

where y is the unique minimal-norm solution to equation $F(u) = 0$.

Theorem 12.1.2. *Under the assumption of Theorem 12.1.1, let $a = a(t)$. Let us assume that*

$$\begin{aligned} 0 < a(t) \searrow 0, \quad \dot a \le 0, \quad \ddot a \ge 0; \quad \lim\nolimits_{t\to\infty} ta(t) = \infty; \\ \lim\nolimits_{t\to\infty} \tfrac{\dot a}{a} = 0; \quad \int_0^\infty a(s)ds = \infty. \end{aligned} \tag{12.1.40}$$

$$\int_0^t e^{-\int_s^t a(\tau)d\tau} |\dot a(s)| ds = O\left(\frac{1}{t}\right) \qquad \text{as } t \to \infty.$$

Then the problem

$$\dot u = -F(u) - a(t)u, \quad u(0) = u_0 \tag{12.1.41}$$

has a unique global solution and

$$\lim_{t\to\infty} ||u(t) - y|| = 0, \tag{12.1.42}$$

where $u(t)$ solves (12.1.41).

In Section 12.2 proofs of Theorems 12.1.1 and 12.1.2 are given.

Remark 12.1.1. Conditions (12.1.40) are satisfied, for example, by the function

$$a(t) = \frac{c_1}{(c_0 + t)^b}, \tag{12.1.43}$$

where c_1, c_0, $b > 0$ are constants, $b \in (0, 1)$. The slow decay of $a(t)$ is important only for large t. For small t the function $a(t)$ may decay arbitrarily fast.

12.2 Proofs

Proof of Theorem 12.1.1. The global existence of the solution $u_a(t)$ to problem (12.1.8) has been proved in Lemma 12.1.3. Let us prove the existence of the limit $u_a(\infty)$. Denote

$$w := u_a(t+h) - u_a(t), \quad ||w(t)|| = g(t), \quad h = \text{const} > 0.$$

Then (12.1.8) implies

$$\dot{w} = -F(u_a(t+h)) - F(u_a(t)) - aw, \quad w(0) = u(h) - u(0). \quad (12.2.1)$$

Multiply (12.2.1) by w and use the monotonicity of F to get

$$g\dot{g} \leq -ag^2, \quad g(0) = ||u(h) - u(0)||. \quad (12.2.2)$$

Thus

$$||u_a(t+h) - u_a(t)|| \leq e^{-at}||u_a(h) - u_a(0)||. \quad (12.2.3)$$

From (12.2.3), (12.1.32), and the Cauchy criterion for the existence of the limit as $t \to \infty$, we conclude that the limit

$$\lim_{t\to\infty} u_a(t) = u_a \quad (12.2.4)$$

exists. Let us denote

$$z_h(t) := \frac{u_a(t+h) - u_a(t)}{h}, \qquad \psi_h(t) := ||z_h(t)||.$$

Then

$$\dot{z}_h = -\frac{1}{h}[F(u_a(t+h)) - F(u_a(t))] - az_h. \quad (12.2.5)$$

Multiply (12.2.5) by z_h and use the monotonicity of F to get

$$\dot{\psi}_h\psi_h \leq -a\psi_h^2. \quad (12.2.6)$$

Thus

$$\psi_h(t) \leq \psi_h(0)e^{-at}. \quad (12.2.7)$$

Let $h \to 0$ in (12.2.7). Since we assumed F continuous, the derivative $\dot{u}(t)$ exists in the strong sense, as the limit

$$\dot{u}(t) = \lim_{h\to 0} z_h(t),$$

and one has:

$$\lim_{h\to 0} \psi_h(t) = ||\dot{u}(t)||.$$

Thus, as $h \to 0$, formula (12.2.7) yields

$$||\dot{u}_a(t)|| \le ||\dot{u}_a(0)||e^{-at}. \tag{12.2.8}$$

This inequality implies the existence of $u(\infty)$ and the following estimates:

$$||u_a(t) - u_a(0)|| \le ||\dot{u}_a(0)||\frac{1-e^{-at}}{a}, \tag{12.2.9}$$

and

$$||u_a(t) - u_a(\infty)|| \le ||\dot{u}_a(0)||\frac{e^{-at}}{a}. \tag{12.2.10}$$

Estimate (12.2.8), the existence of $u(\infty)$ and the continuity of F, allow us to pass to the limit $t \to \infty$ in equation (12.1.8) and get

$$F(u_a(\infty)) + au_a(\infty) = 0. \tag{12.2.11}$$

Let us prove that

$$\lim_{a\to 0} u_a(\infty) = y, \tag{12.2.12}$$

where y is the unique minimal-norm solution to the equation

$$F(y) = 0. \tag{12.2.13}$$

The proof of (12.2.12) is the same as the proof of Lemma 6.1.7: this proof is valid for a hemicontinuous monotone operator F.

Theorem 12.1.1 is proved. □

Proof of Theorem 12.1.2. The global existence of the unique solution to problem (12.1.41) follows from Lemma 12.1.3. Let us prove relation (12.1.41). Equation (12.1.41) and the existence of a solution to (12.2.13) imply:

$$\sup_{t\ge 0} ||u(t)|| < c. \tag{12.2.14}$$

Indeed, let

$$p(t) := u(t) - y, \quad ||p(t)|| = q(t).$$

Equation (12.1.41) implies

$$\dot{p} = -[F(u) - F(y)] - a(t)p - a(t)y. \tag{12.2.15}$$

Multiply this equation by p and use the monotonicity of F to get

$$q\dot{q} \leq -a(t)q^2 + a(t)||y||q,$$

and

$$\dot{q} \leq -a(t)q(t) + a(t)||y||. \tag{12.2.16}$$

This implies

$$\begin{aligned} q(t) &\leq e^{-\int_0^t a(s)ds}\left[q(0) + \int_0^t e^{\int_0^s a(\tau)d\tau} a(s)ds||y||\right] \\ &\leq ||q(0)|| + ||y||, \end{aligned} \tag{12.2.17}$$

because

$$\int_0^t e^{-\int_s^t a(\tau)d\tau} a(s)ds = 1 - e^{-\int_0^t a(\tau)d\tau} \leq 1.$$

Therefore

$$||u(t)|| \leq q(t) + ||y|| \leq ||q(0)|| + 2||y|| := c, \tag{12.2.18}$$

so (12.2.14) is established.

Let us now prove that $q(\infty) = 0$, i.e., the existence of $u(\infty)$ and the relation $u(\infty) = y$.

Using (12.1.41) we obtain

$$\dot{w} = -[F(u(t+h)) - F(u(t))] - [a(t+h)u(t+h) - a(t)u(t)],$$

where

$$w := u(t+h) - u(t).$$

Multiplying this equation by w and using the monotonicity of F, we derive:

$$g\dot{g} \leq |a(t+h) - a(t)|\ ||u(t+h)||g - a(t)g^2, \tag{12.2.19}$$

where

$$g(t) := ||w(t)||,$$

and

$$\dot{g} \leq -a(t)g(t) + ch|\dot{a}(t)|. \qquad (12.2.20)$$

Here we have used estimate (12.2.14) and the assumptions $\dot{a} \leq 0$ and $\ddot{a} > 0$, which imply

$$|a(t+h) - a(t)| \leq h|\dot{a}(t)|.$$

Inequality (12.2.20) implies

$$g(t) \leq e^{-\int_0^t a(s)ds} \left[g(0) + ch \int_0^t e^{\int_0^s a(\tau)d\tau} |\dot{a}(s)| ds \right]. \qquad (12.2.21)$$

Let us derive from estimate (12.2.21) and assumptions (12.1.40) that

$$\lim_{t\to\infty} g(t) = 0 \qquad (12.2.22)$$

uniformly with respect to h running through any fixed compact subset of $\mathbb{R}_+ := [0, \infty)$.

Indeed, the last assumption (12.1.40) implies

$$\lim_{t\to\infty} g(0)e^{-\int_0^t a(s)ds} = 0, \qquad (12.2.23)$$

and an application of the L'Hospital rule yields

$$\lim_{t\to\infty} \frac{\int_0^t e^{\int_0^s a(\tau)d\tau} |\dot{a}(s)| ds}{e^{\int_0^t a(s)ds}} = \lim_{t\to\infty} \frac{|\dot{a}(t)|}{a(t)} = 0, \qquad (12.2.24)$$

because of (12.1.40). Thus, the relation (12.2.22) is established.

From (12.2.14) it follows that there exists a sequence $t_n \to \infty$ such that

$$u(t_n) \rightharpoonup v. \qquad (12.2.25)$$

We want to pass to the limit $t_n \to \infty$ in equation (12.1.41) and obtain $F(v) = 0$. To do this, we note that estimate (12.2.14) and the property $\lim_{t\to\infty} a(t) = 0$ imply

$$\lim_{t\to\infty} a(t)u(t) = 0. \qquad (12.2.26)$$

Dividing inequality (12.2.21) by h and letting $h \to 0$, we obtain

$$||\dot{u}(t)|| \leq e^{-\int_0^t a(s)ds} \left[||\dot{u}(0)|| + c \int_0^t e^{\int_0^s a(\tau)d\tau} |\dot{a}(s)| ds \right]. \qquad (12.2.27)$$

This inequality together with (12.2.23) and (12.2.24) implies

$$\lim_{t\to\infty} ||\dot{u}(t)|| = 0. \tag{12.2.28}$$

From (12.2.26), (12.2.28) and (12.1.41) it follows that

$$\lim_{t\to\infty} F(u(t)) = 0. \tag{12.2.29}$$

From (12.2.25), (12.2.29) and Lemma 6.1.4 it follows that

$$F(v) = 0. \tag{12.2.30}$$

Let us prove now that

$$\lim_{n\to\infty} u(t_n) = v \quad \text{and} \quad v = y. \tag{12.2.31}$$

From (12.2.25) we conclude that

$$||v|| \le \liminf_{n\to\infty} ||u(t_n)||. \tag{12.2.32}$$

Let us prove that

$$\limsup_{n\to\infty} ||u(t_n)|| \le ||v||. \tag{12.2.33}$$

If (12.2.33) is proved, then (12.2.32) and (12.2.33) imply

$$\lim_{n\to\infty} ||u(t_n)|| = ||v||. \tag{12.2.34}$$

This relation together with weak convergence (12.2.25) imply strong convergence (12.2.31).

So, *let us verify (12.2.33)*. By the last assumption (12.1.40) we have

$$e^{-\int_0^t a(s)ds} \int_0^t e^{\int_0^s a(\tau)d\tau} |\dot{a}(s)| ds = O\left(\frac{1}{t}\right) \quad \text{as} \quad t \to \infty. \tag{12.2.35}$$

Thus, estimate (12.2.27) for $a(t)$, defined in (12.1.43), implies

$$||\dot{u}(t)|| = O\left(\frac{1}{t}\right) \quad \text{as} \quad t \to \infty. \tag{12.2.36}$$

Equation (12.2.30) and (12.1.41) imply:

$$(F(u(t_n)) - F(v), u(t_n) - v) + a(t_n)(u(t_n), u(t_n) - v) = -(\dot{u}(t_n), u(t_n) - v). \tag{12.2.37}$$

This relation holds for any solution to equation (12.2.30). Equation (12.2.37) and the monotonicity of F imply

$$(u(t_n), u(t_n) - v) \leq -\frac{1}{a(t_n)}(\dot{u}(t_n), u(t_n) - v). \tag{12.2.38}$$

From (12.2.14), (12.2.36) and (12.2.38) we get

$$(u(t_n), u(t_n) - v) \leq \frac{c}{t_n a(t_n)} \to 0 \quad \text{as} \quad n \to \infty, \tag{12.2.39}$$

because

$$\lim_{t\to\infty} ta(t) = \infty$$

by the assumption (12.1.40). From (12.2.39) we obtain (12.2.33). Therefore, the first relation (12.2.31) is verified.

Let us prove that $v = y$. In formula (12.2.37) we can replace v by any solution to equation (12.2.30). Let us replace v by y. Then (12.2.39) is replaced by

$$||u(t_n)||^2 \leq ||u(t_n)||||y|| + \frac{c}{t_n a(t_n)}. \tag{12.2.40}$$

Passing to the limit $n \to \infty$, yields

$$\limsup_{n\to\infty} ||u(t_n)|| \leq ||y||. \tag{12.2.41}$$

We have already proved that

$$\limsup_{n\to\infty} ||u(t_n)|| = \lim_{n\to\infty} ||u(t_n)|| = ||v||. \tag{12.2.42}$$

Thus

$$||v|| = ||y||. \tag{12.2.43}$$

Since y is the unique minimum-norm solution of equation (12.2.30) and v solves this equation and has a norm greater than $||y||$, it follows that $v = y$.

Theorem 12.1.2 is proved. □

Chapter 13

DSM as a theoretical tool

In this Chapter we give a sufficient condition for a nonlinear map to be surjective and to be a global homeomorphism.

13.1 Surjectivity of nonlinear maps

In this Section we prove the following result.

Theorem 13.1.1. *Assume that $F : X \to X$ is a C^2_{loc} map in a Banach space X and assumptions (1.3.1), (1.3.2) hold. Then F is surjective if*

$$\limsup_{R\to\infty} \frac{R}{m(R)} = \infty. \tag{13.1.1}$$

Proof. Consider the DSM:

$$\dot{u} = -[F'(u)]^{-1}(F(u) - f), \quad u(0) = u_0, \tag{13.1.2}$$

where f and u_0 are arbitrary. Arguing as in the proof of Theorem 3.2.1 we establish the existence and uniqueness of the global solution to (1.3.2) provided that condition analogous to (3.2.6) holds:

$$||F(u_0) - f||m(R) \le R. \tag{13.1.3}$$

If this condition holds, then $u(\infty)$ exists and

$$F(u(\infty)) = f. \tag{13.1.4}$$

Since f is arbitrary, the map F is surjective. Our assumption (13.1.1) implies that for sufficiently large R and an arbitrary fixed u_0 condition (13.1.3) is satisfied.

Theorem 13.1.1 is proved. □

13.2 When is a local homeomorphism a global one?

Let $F: X \to X$ satisfy conditions (1.3.1) and (1.3.2). Condition (1.3.1) implies that F is a local homeomorphism, i.e., F maps a sufficiently small neighborhood of an arbitrary point u_0 onto a small neighborhood of the point $F(u_0)$ bijectively and bicontinuously.

It is well-known that a local homeomorphism may be not a global one. See an example in Section 2.6. Here is another one.

Example 13.2.1. Let

$$F : \mathbb{R}^2 \to \mathbb{R}^2, \qquad F(u) = \begin{pmatrix} \arctan u_1 \\ u_2(1+u_1^2) \end{pmatrix}.$$

Then $\det F'(u) = 1$, but F is not surjective. Indeed

$$F'(u) = \begin{pmatrix} \frac{1}{1+u_1^2} & 0 \\ 2u_1u_2 & 1+u_1^2 \end{pmatrix}, \qquad \det F'(u) = 1,$$

so

$$[F'(u)]^{-1} = \begin{pmatrix} 1+u_1^2 & 0 \\ -2u_1u_2 & \frac{1}{1+u_1^2} \end{pmatrix}.$$

There is no point $u = \begin{pmatrix} u_1 \\ u_2 \end{pmatrix}$ such that $F(u) = \begin{pmatrix} \frac{\pi}{2} \\ 1 \end{pmatrix}$.

This example can be found in [OR].

J. Hadamard has proved in 1906 (see [Ha]) the following result:

Proposition 13.2.1. *If*

$$F : \mathbb{R}^n \to \mathbb{R}^n, \quad F \in C^1_{loc}(\mathbb{R}^n), \quad \textit{and} \quad \sup_{u \in \mathbb{R}^n} ||[F'(u)]^{-1}|| \le m, \tag{13.2.1}$$

then F is a global homeomorphism of $\mathbb{R}^n$ onto $\mathbb{R}^n$.

This result was generalized later to Hilbert and Banach spaces (see [OR] and references therein).

Our aim is to give the following generalization of the Hadamard's result.

Theorem 13.2.2. *Assume that $F : H \to H$ satisfies assumption (1.3.2) and*

$$||[F'(u)]^{-1}|| \le \psi(||u||), \tag{13.2.2}$$

where H is a Hilbert space and $\psi(s) > 0$ is a continuous function on $[0, \infty)$ such that

$$\int_0^\infty \frac{ds}{\psi(s)} = \infty. \tag{13.2.3}$$

Then F is a global homeomorphism of H onto H.

Remark 13.2.1. One can take, for example, $\psi(s) = m = \text{const}$, and then the assumption (13.2.2) reduces to (13.2.1). One can take $\psi(s) = as + b$, where $a, b > 0$ are constants.

Proof of Theorem 13.2.2. It follows from (13.2.2) that F is a local homeomorphism. We want to prove that F is injective and surjective.

Let us first prove that (13.2.2) implies surjectivity of F.

Consider problem (13.1.2). Denote

$$||F(u(t)) - f|| := g(t). \tag{13.2.4}$$

Under the assumptions of Theorem 13.2.2 problem (13.1.2) has a unique local solution. We will establish a uniform with respect to t bound

$$\sup_t ||u(t)|| \le c, \tag{13.2.5}$$

and then the local solution to (13.1.2) is a global one (see Lemma 2.6.1). Then we prove the existence of $u(\infty)$ and that $u(\infty)$ solves equation (13.1.4). This will prove the surjectivity of F.

We have, using (13.1.2),

$$g\dot{g} = \text{Re}(F'(u)\dot{u}, F(u) - f) = -g^2.$$

Thus

$$g(t) = g(0)e^{-t}. \tag{13.2.6}$$

This and (13.1.2) imply

$$||\dot{u}|| \le \psi(||u||)g(0)e^{-t}. \tag{13.2.7}$$

Note that

$$||u||^{\cdot} \le ||\dot{u}||. \tag{13.2.8}$$

Indeed, differentiate $(u, u) = |u||^2$ with respect to t and get

$$2||u||||u||^{\cdot} = 2\text{Re}(\dot{u}, u) \le 2||\dot{u}||||u||.$$

If $||u(t)|| > 0$ then we get (13.2.8). If the closure of the set of points t at which $u(t) = 0$ contains an open set, then on this set $||u||^{\cdot} = ||\dot{u}|| = 0$, so (13.2.8) holds. If inequality (13.2.8) holds everywhere except on a nowhere dense set, then by continuity of $||\dot{u}(t)||$ it holds everywhere.

Thus, denoting $||u(t)|| := s$, we can derive from (13.2.7) the inequality

$$\int_{||u_0||}^{||u(t)||} \frac{ds}{\psi(s)} \leq g(0)(1 - e^{-t}). \tag{13.2.9}$$

Thus inequality and assumption (13.2.3) imply estimate (13.2.5). From (13.2.5) and (13.2.7) it follows that

$$||\dot{u}(t)|| \leq c_1 e^{-t}, \tag{13.2.10}$$

where

$$c_1 := g(0) \max_{0 \leq s \leq c} \psi(s).$$

Estimate (13.2.10) implies the existence of $u(\infty)$ and the inequalities

$$||u(t) - u(\infty)|| \leq c_1 e^{-t}, \quad ||u(t) - u_0|| \leq c_1. \tag{13.2.11}$$

Passing to the limit $t \to \infty$ in equation (13.1.2) and taking into account (13.2.10) we derive equation

$$F(v) - f = 0, \quad v := u(\infty). \tag{13.2.12}$$

Since f is arbitrary, the map F is surjective.

Let us now prove that F is injective, i.e., $F(v) = F(w)$ implies $w = v$. The idea of our proof is simple. We started with the initial approximation u_0 in problem (13.1.2) and constructed $v = u(\infty; u_0)$. We wish to consider the straight line

$$u_0(s) = u_0 + s(w - u_0), \quad u_0(0) = u_0, \quad u_0(1) = w, \tag{13.2.13}$$

and for each $u_0(s)$ we construct

$$u(t, s) := u(t, u_0(s)).$$

We will show that

$$u(\infty, s) = w \quad \forall s \in [0, 1]. \tag{13.2.14}$$

This implies $w = v$.

To verify (13.2.14) we show that if

$$||u(\infty, s) - u(\infty, s+\sigma)||$$

is sufficiently small, then the equation

$$F(u(\infty, s)) = F(u(\infty, s+\sigma)) = f$$

implies

$$u(\infty, s) = u(\infty, s+\sigma), \tag{13.2.15}$$

because F is a local homeomorphism.

We prove that

$$||u(\infty, s) - u(\infty, s+\sigma)|| \leq c||u(0, s) - u(0, s+\sigma)|| \leq c_2\sigma, \tag{13.2.16}$$

where

$$c_2 := c||w - u_0||,$$

so that $||u(\infty, s) - u(\infty, s+\sigma)||$ is as small as we wish if σ is sufficiently small. Thus, in finitely many steps we can reach the point $s = 1$, and at each step we will have equation (13.2.14). So, our proof will be complete if we verify estimate (13.2.16).

Let us do this. Denote

$$x(t) := u(t, s+\sigma) - u(t, s), \quad \eta(t) := ||x(t)||, \tag{13.2.17}$$

$$z := u(t, s+\sigma), \quad \zeta := u(t, s), \quad z - \zeta = x = x(t). \tag{13.2.18}$$

Using (13.2.10), (13.1.2) and (1.3.2), we get:

$$\begin{aligned} \eta\dot{\eta} &= -([F'(z)]^{-1}(F(z) - f) - [F'(\zeta)]^{-1}(F(\zeta) - f), x(t)) \\ &= (([F'(z)]^{-1} - [F'(\zeta)]^{-1})(F(z) - f), x(t)) \\ &\quad -([F'(z)]^{-1}(F(z) - F(\zeta)), x(t)) \\ &\leq ce^{-t}\eta^2 - \eta^2 + c\eta^3, \quad \eta \geq 0. \end{aligned} \tag{13.2.19}$$

Here we have used the following estimates:

$$||[F'(z)]^{-1} - [F'(\zeta)]^{-1}|| \leq c||z - \zeta||, \tag{13.2.20}$$

$$||F(z) - f|| \leq ce^{-t}, \tag{13.2.21}$$

$$F(z) - F(\zeta) = F'(\zeta)(z - \zeta) + K, \quad ||K|| \leq \frac{M_2\eta^2(t)}{2}, \tag{13.2.22}$$

$$||[F'(\zeta)]^{-1}|| \le c, \tag{13.2.23}$$

where $c > 0$ denotes different constants independent of time. From equation (13.2.19) we obtain the following inequality:

$$\dot{\eta} \le -\eta + ce^{-t}\eta + c\eta^2, \quad \eta(0) = ||u(0, s + \sigma) - u(0, s)|| := \delta. \tag{13.2.24}$$

Let us define $q = q(t)$ by formula:

$$\eta = q(t)e^{-t}. \tag{13.2.25}$$

Then (13.2.24) yields:

$$\dot{q} \le ce^{-t}(q + q^2), \quad q(0) = \delta. \tag{13.2.26}$$

We integrate (13.2.26) and get

$$\int_{\delta}^{q(t)} \frac{dq}{q + q^2} \le c(1 - e^{-t}). \tag{13.2.27}$$

This implies

$$\ln \frac{q(\delta + 1)}{(q + 1)\delta} \le c, \quad \frac{q}{q + 1} \le \frac{\delta}{\delta + 1}e^c := c_2\delta, \tag{13.2.28}$$

where $c_2 = \frac{e^c}{\delta+1}$. We assume that

$$c_2\delta < 1. \tag{13.2.29}$$

Then (13.2.28) yields

$$q(t) \le c_3\delta. \tag{13.2.30}$$

Therefore

$$\begin{aligned} ||u(t, s + \sigma) - u(t, s)|| &:= \eta(t) \\ &\le c_3\delta e^{-t}||u(0, s + \sigma) - u(0, s)|| \le c_4 e^{-t}\sigma. \end{aligned} \tag{13.2.31}$$

We have verified estimate (13.2.16). This completes the proof of injectivity of F.

Theorem 13.2.2 is proved. □

Chapter 14

DSM and iterative methods

14.1 Introduction

In this Chapter a general approach to constructing convergent iterative schemes is developed. This approach is based on the DSM. The idea is simple. Suppose we want to solve an operator equation $F(u) = 0$ which has a solution, and we have justified a DSM method

$$\dot{u} = \Phi(t, u), \quad u(0) = u_0 \tag{14.1.1}$$

for solving equation $F(u) = 0$. Suppose also that a dicretization scheme

$$u_{n+1} = u_n + h_n \Phi(t_n, u_n), \quad u_0 = u_0, \quad t_{n+1} = t_n + h, \tag{14.1.2}$$

converges to the solution of (14.1.1) and one can choose h_n so that

$$\sum_{n=1}^{\infty} h_n = t,$$

where $t \subset \mathbb{R}_+$ is arbitrary, then (14.1.2) is a convergent iterative process for solving equation $F(u) = 0$.

Iterative methods for solving linear equations have been discussed in Section 2.4 and 4.4. The basic result we have proved in these Sections can be described as follows:

Every solvable linear equation $Au = f$ with a closed densely defined operator in a Hilbert space H can be solved by a convergent iterative process.

If the data f_δ are given, such that $||f_\delta - f|| \le \delta$, then the iterative process gives a stable approximation to the minimal-norm solution of the equation $Au = f$ provided that iterations are stopped at $n = n(\delta)$, where $n(\delta)$ is suitably chosen.

In Section 14.2 we construct an iterative method for solving well-posed problems. We prove that every solvable well-posed operator equation can be solved by an iterative method which converges at an exponential rate. In Section 14.3 we construct an iterative process for solving ill-posed problems with monotone operators. We prove that any solvable nonlinear operator equation $F(u) = f$ with C^2_{loc} monotone operator F can be solved by an iterative process which converges to the unique minimal-norm solution y of this equation for any choice of the initial approximation. In Section 14.4 we deal with iterative processes for general nonlinear equations.

14.2 Iterative solution of well-posed problems

Consider the equation

$$F(u) = 0, \tag{14.2.1}$$

where $F : H \to H$ satisfies conditions (1.3.1) and (1.3.2) and H is a Hilbert space. Let a DSM for solving equation (14.2.1) be of the form

$$\dot{u} = \Phi(u), \quad u(0) = u_0, \tag{14.2.2}$$

and assume that conditions (3.1.23) and (3.1.24) hold. Note that in Sections 3.1 - 3.6 the Φ did not depend on t. That is why we take $\Phi(u)$ rather than $\Phi(t, u)$ in (14.2.2). We assume that Φ is locally Lipschitz.

Consider the following iterative process

$$u_{n+1} = u_n + h\Phi(u_n), \quad u_0 = u_0, \tag{14.2.3}$$

where the initial approximation u_0 is chosen so that condition (3.1.25) hold.

Theorem 14.2.1. *If F satisfies conditions (1.3.1), (1.3.2), (3.1.25) and Φ satisfies conditions (3.1.23) and (3.1.24), and if $h > 0$ in (14.2.3) is sufficiently small, then iterative process (14.2.3) produces u_n such that*

$$||u_n - y|| \leq Re^{-chn}, \quad ||F(u_n)|| \leq ||F_0||e^{-chn}, \tag{14.2.4}$$

where

$$F_0 := F(u_0), \quad c = const > 0, \quad c < c_1,$$

c_1 is the constant from condition (3.1.23), $R > 0$ is a constant from conditions (1.3.1) and (1.3.2), and y solves equation (14.2.1).

Proof. For $n = 0$ the second estimate (14.2.4) is obvious, and the first one follows from (3.1.27) and (3.1.25). Assuming that estimates (14.2.4) hold for $n \le m$, let us prove that they hold for $n = m + 1$. Then, by induction, they hold for any n. Denote by $w_{n+1}(t)$ the solution to the problem

$$\dot{w}(t) = \Phi(w), \quad w(nh) = u_n, \quad t_n := nh \le t \le (n+1)h =: t_{n+1}. \tag{14.2.5}$$

Estimate (3.1.26) yields

$$||w(t) - y|| \le \frac{c_2}{c_1} ||F_n|| e^{-c_1 t} \le R e^{-chn - c_1(t-nh)}, \quad t \ge t_n. \tag{14.2.6}$$

We have

$$||u_{n+1} - y|| \le ||u_{n+1} - w(t_{n+1})|| + ||w(t_{n+1}) - y||, \tag{14.2.7}$$

and

$$\begin{aligned} ||u_{n+1} - w(t_{n+1})|| &\le \int_{t_n}^{t_{n+1}} ||\Phi(u_n) - \Phi(w(s))|| ds \\ &\le L_1 \int_{t_n}^{t_{n+1}} ||u_n - w(s)|| ds \\ &\le L_1 \int_{t_n}^{t_{n+1}} ds || \int_{t_n}^{s} \Phi(w(s)) ds|| \\ &\le L_1 h c_2 \int_{t_n}^{t_{n+1}} ||F(w(s))|| ds \\ &\le L_1 c_2 h^2 ||F(u_n)|| \\ &\le L_1 c_2 h^2 ||F_0|| e^{-chn}. \end{aligned} \tag{14.2.8}$$

From (14.2.7), (14.2.8) and (14.2.6) with $t = (n+1)h$ we obtain

$$||u_{n+1} - y|| \le R e^{-chn} \left(e^{-c_1 h} + L_1 c_1 h^2 \right) \le R e^{-ch(n+1)}, \tag{14.2.9}$$

provided that

$$e^{-c_1 h} + L_1 c_1 h^2 \le e^{-ch}. \tag{14.2.10}$$

Inequality (14.2.10) holds if $c_1 > c$ and h is sufficiently small.

Let us estimate $||F(u_{n+1})||$. We have

$$||F(u_{n+1})|| \le ||F(u_{n+1}) - F(w(t_{n+1}))|| + ||F(w(t_{n+1}))||. \tag{14.2.11}$$

Furthermore, using (14.2.8) and (1.3.2) we get

$$||F(u_{n+1}) - F(w(t_{n+1}))|| \le M_1 ||u_{n+1} - w(t_{n+1})|| \le M_1 L_1 c_2 h^2 ||F_0|| e^{-chn}$$

(14.2.12)

and using (3.1.28) with $u_0 = u_n$ and $t = t_{n+1} - t_n = h$, we get

$$||F(w(t_{n+1}))|| \le ||F(u_n)||e^{-c_1 h} \le ||F_0||e^{-chn-c_1 h}. \quad (14.2.13)$$

From (14.2.11) - (14.2.13) we obtain

$$||F(u_{n+1})|| \le ||F_0||e^{-chn}\left(e^{-c_1 h} + M_1 L_1 c_2 h^2\right) \le ||F_0||e^{-ch(n+1)}, \quad (14.2.14)$$

provided that

$$e^{-c_1 h} + M_1 L_1 c_2 h^2 \le e^{-ch}. \quad (14.2.15)$$

This inequality holds if $c_1 > c$ and h is sufficiently small.

Finally, the assumption (3.1.25) implies the existence of a solution y to equation (14.2.1).

Theorem 14.2.1 is proved. □

Remark 14.2.1. If condition (3.1.25) is dropped but we assume that equation (14.2.1) has a solution y and $||u_0 - y||$ is sufficiently small, then our proof remains valid and yields the conclusion (14.2.4) of Theorem 14.2.1.

14.3 Iterative solution of ill-posed equations with monotone operator

Assume now that equation (14.2.1) has a solution y, the operator F in this equation satisfies assumption (1.3.2) but not (1.3.1), and F is monotone:

$$(F(u) - F(v), u - v) \ge 0, \quad \forall u, v \in H, \quad F : H \to H, \quad (14.3.1)$$

where H is a Hilbert space. Let y be the (unique) minimal-norm solution to equation (14.2.1).

Theorem 14.3.1. *Under the above assumptions one can choose $h_n > 0$ and $a_n > 0$ such that the iterative process*

$$u_{n+1} = u_n - h_n A_n^{-1}[F(u_n) + a_n u_n], \quad u_0 = u_0, \quad (14.3.2)$$

where $A_n := F'(u_n) + a_n I$, and $u_0 \in H$ is arbitrary, converges to y:

$$\lim_{n\to\infty} ||u_n - y|| = 0. \quad (14.3.3)$$

Proof. Denote by V_n the solution of the equation

$$F(V_n) + a_n V_n = 0. \tag{14.3.4}$$

This equation has a unique solution as we have proved in Lemma 6.1.6. Let

$$z_n := u_n - V_n, \quad ||z_n|| := g_n. \tag{14.3.5}$$

We have

$$||u_n - y|| \leq g_n + ||V_n - y||. \tag{14.3.6}$$

In Lemma 6.1.7 we have proved that

$$\lim_{n\to\infty} ||V_n - y|| = 0, \tag{14.3.7}$$

provided that

$$\lim_{n\to\infty} a_n = 0. \tag{14.3.8}$$

Let us prove that

$$\lim_{n\to\infty} ||u_n - V_n|| = 0. \tag{14.3.9}$$

Denote

$$b_n := ||V_{n+1} - V_n||.$$

Then

$$\lim_{n\to\infty} b_n = 0.$$

Let us write (14.2.1) as

$$z_{n+1} = (1 - h_n)z_n - h_n A_n^{-1} K(z_n) - (V_{n+1} - V_n). \tag{14.3.10}$$

Here we have used Taylor formula:

$$F(u_n) + a_n u_n = F(v_n) - F(V_n) + a_n(u_n - V_n) = A_n z_n + K(z_n), \tag{14.3.11}$$

where

$$||K(z_n)|| \leq \frac{M_2}{2} ||z_n||^2 := c g_n^2. \tag{14.3.12}$$

From (14.3.10) we obtain

$$g_{n+1} \le (1-h_n)g_n + \frac{ch_n}{a_n} g_n^2 + b_n, \quad 0 < h_n \le 1. := (1-\gamma_n)g_n + b_n, \quad (14.3.13)$$

where the estimate

$$||A_n^{-1}|| \le \frac{1}{a_n}$$

is used. Choose

$$a_n = 2cg_n. \tag{14.3.14}$$

Then

$$\begin{aligned} g_{n+1} &\le (1-h_n)g_n + \frac{h_n}{2} g_n + b_n \\ &= \left(1 - \frac{h_n}{2}\right) g_n + b_n := (1-\gamma_n)g_n + b_n. \end{aligned}$$

Therefore

$$g_{n+1} \le (1-\gamma_n)g_n + b_n, \tag{14.3.15}$$

where

$$\gamma_n := \frac{h_n}{2}, \quad 0 < \gamma_n \le \frac{1}{2},$$

and $g_1 \ge 0$ is arbitrary.

Assume that

$$\sum_{n=1}^{\infty} h_n = \infty. \tag{14.3.16}$$

Then (14.3.15) implies

$$\lim_{n\to\infty} g_n = 0. \tag{14.3.17}$$

Indeed, (14.3.15) implies:

$$g_{n+1} \le b_n + \sum_{k=1}^{n-1} b_k \prod_{j=k+1}^{n} (1-\gamma_j) + g_1 \prod_{j=1}^{n} (1-\gamma_j). \tag{14.3.18}$$

Assumption (14.3.16) implies

$$\lim_{n\to\infty} \prod_{j=1}^{n} (1-\gamma_j) = 0. \tag{14.3.19}$$

Assume that (14.3.16) holds and

$$\lim_{n\to\infty} \frac{b_{n-1}}{h_n} = 0. \tag{14.3.20}$$

Then

$$\lim_{n\to\infty} \sum_{k=1}^{n-1} b_k \prod_{j=k+1}^{n} (1-\gamma_j) = 0. \tag{14.3.21}$$

This follows from a discrete analog of L'Hospital's rule:

If $p_n, q_n > 0$, $\lim_{n\to\infty} q_n = \infty$, and $\lim_{n\to\infty} \frac{p_n - q_{n-1}}{q_n - q_{n-1}}$ exists, then

$$\lim_{n\to\infty} \frac{p_n}{q_n} = \lim_{n\to\infty} \frac{p_n - p_{n-1}}{q_n - q_{n-1}}.$$

In our case

$$p_n = \sum_{k=1}^{n-1} b_k \prod_{j=1}^{k} (1-\gamma_j), \quad q_n = \prod_{j=1}^{n} (1-\gamma_j)^n, \quad \lim_{n\to\infty} q_n = \infty,$$

$$\begin{aligned} \lim_{n\to\infty} \frac{p_n - p_{n-1}}{q_n - q_{n-1}} &= \lim_{n\to\infty} \frac{b_{n-1} \prod_{j=1}^{n-1} (1-\gamma_j)}{\prod_{j=1}^{n} (1-\gamma_j) - \prod_{j=1}^{n-1} (1-\gamma_j)} \\ &= \lim_{n\to\infty} \frac{b_{n-1}}{-\gamma_n} = 0, \end{aligned} \tag{14.3.22}$$

where we have used assumption (14.3.20).

Theorem 14.3.1 is proved.

□

Remark 14.3.1. If $h = \text{const}$, $h \in (0,1)$, then condition (14.3.16) holds, so one can use $h_n = h$, $h \in (0,1)$, in the iterative process (14.3.2) with a_n chosen as in (14.3.14). Then (14.3.15) takes the form

$$g_{n+1} \le q g_n + b_n, \quad q := 1 - h \in (0,1), \quad \lim_{n\to\infty} b_n = 0, \tag{14.3.23}$$

g_1 is arbitrary, and (14.3.23) implies $\lim_{n\to\infty} g_n = 0$.

14.4 Iterative methods for solving nonlinear equations

In this Section we do not assume that $F : H \to H$ is monotone. We assume that condition (1.3.2) holds, that y solves equation (14.2.1), $F(y) = 0$, and that

$$\tilde{A} := F'(y) \neq 0. \tag{14.4.1}$$

We want to construct a convergent iterative process for solving equation (14.2.1).

The DSM for solving this equation has been given in Section 7.1, formula (7.1.3), and we use the notations and the results from this Section. Consider the corresponding iterative method:

$$u_{n+1} = u_n - h_n T_{a_n}^{-1}[A^*(u_n)F(u_n) + a_n(u_n - z)], \quad u_0 = u_0, \tag{14.4.2}$$

where $a_n, h_n > 0$ are some sequences,

$$T := A^*(u_n)A(u_n), \quad T_{a_n} := T + a_n I.$$

Following the method developed in Section 7.1, let us denote

$$u_n - u := w_n, \quad ||w_n|| = g_n, \tag{14.4.3}$$

and rewrite (14.4.2) as follows:

$$w_{n+1} = w_n - h_n T_{a_n}^{-1}[A^*(u_n)(F(u_n) - F(y)) + a_n w_n + a_n(y - z)]. \tag{14.4.4}$$

As was proved in Section 7.1, we can choose z so that

$$y - z = \tilde{T}v, \quad 2M_1 M_2 ||v|| \leq \frac{1}{2}, \tag{14.4.5}$$

where M_1 and M_2 are constants from (1.3.2). This is possible if assumption (14.4.1) holds.

Let us use the formulas:

$$F(u_n) - F(y) = A(u_n)w_n + K(w_n), \quad ||K(w_n)|| \leq \frac{M_2}{2} g_n^2, \tag{14.4.6}$$

$$||T_{a_n}^{-1} A^*(u_n)|| \leq \frac{1}{2\sqrt{a_n}}, \tag{14.4.7}$$

and rewrite (14.4.4) as follows:

$$w_{n+1} = (1 - h_n)w_n - h_n T_{a_n}^{-1} A^*(u_n)K(w_n) - h_n a_n T_{a_n}^{-1}\tilde{T}v. \tag{14.4.8}$$

We have

$$T_{a_n}^{-1}\tilde{T} = \left(T_{a_n}^{-1} - \tilde{T}_{a_n}^{-1}\tilde{T}\right)\tilde{T} + \tilde{T}_{a_n}^{-1}\tilde{T}.$$

Therefore

$$||T_{a_n}^{-1}\tilde{T}v|| \leq ||\left(T_{a_n}^{-1} - \tilde{T}_{a_n}^{-1}\right)\tilde{T}||||v|| + ||v||, \tag{14.4.9}$$

where the inequality

$$||\tilde{T}_a^{-1}\tilde{T}|| \leq 1, \quad a \geq 0,$$

was used.

From (14.4.7) - (14.4.9) we obtain

$$\begin{aligned} g_{n+1} \quad &\leq \quad (1-h_n)g_n + h_n \frac{M_2}{4\sqrt{a_n}} g_n^2 + h_n a_n ||v|| \\ &\quad + h_n a_n ||v|| || \left(T_{a_n}^{-1} - \tilde{T}_{a_n}^{-1}\right)\tilde{T}||. \end{aligned} \tag{14.4.10}$$

Denote

$$C_0 := \frac{M_2}{2},$$

and use the following estimate:

$$\begin{aligned} ||\left(T_{a_n}^{-1} - \tilde{T}_{a_n}^{-1}\right)\tilde{T}|| \quad &= \quad ||T_{a_n}^{-1}[A^*(u_n)A(u_n) - \tilde{A}^*(y)\tilde{A}(y)]\tilde{T}_{a_n}^{-1}\tilde{T}|| \\ &\leq \quad \frac{2M_1M_2g_n}{a_n} := \frac{c_1 g_n}{a_n}. \end{aligned} \tag{14.4.11}$$

From (14.4.10) and (14.4.11) we get

$$g_{n+1} \leq \left(1 - \frac{h_n}{2}\right) g_n + \frac{c_0 h_n}{\sqrt{a_n}} g_n^2 + h_n a_n ||v||, \tag{14.4.12}$$

where we have used inequality (14.4.5), which implies

$$c_1||v|| \leq \frac{1}{2}.$$

Let us choose a_n so that

$$\frac{c_0}{\sqrt{a_n}} g_n = \frac{1}{4}, \quad \text{i.e.,} \quad a_n = 16c_0^2 g_n^2. \tag{14.4.13}$$

Then (14.4.12) takes the form

$$g_{n+1} \le \left(1 - \frac{h_n}{4}\right) g_n + 16h_n ||v|| c_0^2 g_n^2, \quad g_0 = ||u_0 y|| \le R, \qquad (14.4.14)$$

where $R > 0$ is the radius of the ball in condition (1.3.2).

We can choose $h_n = h \in (0,1)$ and $g_0 > 0$ such that (14.4.14) implies

$$\lim_{n\to\infty} g_n = 0. \qquad (14.4.15)$$

This possibility is seen from the following lemma.

Lemma 14.4.1. *Let*

$$g_{n+1} \le \gamma g_n + p g_n^2, \quad g_0 = m > 0; \quad 0 < \gamma < 1, \quad p > 0, \qquad (14.4.16)$$

where p is an arbitrary fixed positive constant. If $m > 0$ is sufficiently small, namely, if

$$m < (q - \gamma)/p,$$

where $q \in (\gamma, 1)$ is an arbitrary number, then the sequence g_n, generated by (14.4.16) tends to zero:

$$\lim_{n\to\infty} g_n = 0 \qquad (14.4.17)$$

at a rate of a geometrical progression,

$$g_n \le mq^n, \quad 0 < q < 1.$$

Proof. Assumption (14.4.16) implies

$$g_{n+1} < q^{n+1} m, \quad \gamma < q < 1, \quad n = 0, 1, 2, \cdots \qquad (14.4.18)$$

provided that

$$m < \frac{q - \gamma}{p}, \quad \gamma < q < 1. \qquad (14.4.19)$$

Let us prove (14.4.18) by induction. We have

$$q_1 \le \gamma m + pm^2 < qm, \quad m := g_0,$$

provided (14.4.19) holds. So (14.4.18) holds for $n = 0$. Assuming that it holds for some n, let us check that it holds for $n + 1$:

$$\begin{aligned} g_{n+2} &\le \gamma g_{n+1} + p g_{n+1}^2 \le \gamma q^{n+1} m + p q^{2n+2} m^2 \\ &\le q^{n+1}(\gamma m + pm^2) < q^{n+2} m. \end{aligned} \qquad (14.4.20)$$

So Lemma 14.4.1 is proved. □

From Lemma 14.4.1 it follows that we can choose a constant $h_n = h$ in (14.4.14) such that by choosing g_0 sufficiently small we get (14.4.15).

Let us formulate the result we have proved.

Theorem 14.4.1. *Assume tht $F : F \to H$ satisfies conditions (1.3.2) and (14.4.1), where y solves equation (14.2.1). Then the iterative process (14.4.2) produces a sequence u_n such that*

$$\lim_{n\to\infty} ||u_n - y|| = 0, \tag{14.4.21}$$

provided that a_n is chosen as in (14.3.3) and $h_n = h$ is a constant, $h \in (0, 1)$, $u_0 - y$ is sufficiently small, and z is chosen so that (14.4.5) holds. The convergence in (14.4.21) is at the rate of a geometrical series at least.

Remark 14.4.1. Theorem 14.4.1 guarantees that any solvable equation (14.2.1) can be solved by a convergent iterative process if conditions (1.3.2) and (14.4.1) hold. We do not give an algorithm for choosing z, but only prove the existence of such a z that (14.4.5) holds.

14.5 Ill-posed problems

Suppose that the assumptions of Theorem 14.3.1 or Theorem 14.4.1 hold, that equation

$$F(u) = f \tag{14.5.1}$$

is being considered, and that noisy data f_δ, $||f_\delta - f|| \leq \delta$, are given in place of the exact data f for which equation (14.5.1) is solvable. We want to show that iterative processes (14.3.2) and (14.4.2) can be used for constructing a stable approximation to a solution y of equation (14.5.1). This is done by the method developed in Section 4.4. Namely, we stop iterations at the stopping number $n = n(\delta)$, $\lim_{\delta\to 0} n(\delta) = \infty$, which can be chosen so that $u_\delta := u_{n(\delta)}(f_\delta)$ is the desired stable approximation of y in the sense

$$\lim_{\delta\to 0} ||u_\delta - y|| = 0. \tag{14.5.2}$$

Here $u_{n(\delta)}(f_\delta)$ is the sequence, generated by the iterative process (14.3.2) (or (14.4.2)) with $F(u_n) - f_\delta$ replacing $F(u_n)$. Consider, for example, iterative process (14.3.2), where u_n is replaced by $u_{n,\delta} := u_n(f_\delta)$ and $F(u_n)$ is replaced by $F(u_n) - f_\delta$. We have proved in Theorem 14.3.1 that

$$\lim_{n\to\infty} ||u_n(f) - y|| = 0, \tag{14.5.3}$$

where $u_n := u_n(f)$ is generated by (14.3.2) with $F(u_n) - f_\delta$ replacing $F(u_n)$, f being the exact data. We have

$$||u_n(f_\delta) - y|| \leq ||u_n(f_\delta) - u_n(f)|| + ||u_n(f) - y||. \tag{14.5.4}$$

If we choose $n = n(\delta)$ such that

$$\lim_{\delta \to 0} n(\delta) = \infty \tag{14.5.5}$$

and

$$\lim_{\delta \to 0} ||u_{n(\delta)}(f_\delta) - u_{n(\delta)}(f)|| = 0, \tag{14.5.6}$$

then, due to (14.5.3), the element $u_\delta := u_{n(\delta)}(f_\delta)$ gives the desired stable approximation of the solution y. Due to the continuity of $u_n(f)$ with respect to f for any fixed n, we have:

$$||u_n(f_\delta) - u_n(f)|| \leq \epsilon(n, \delta), \tag{14.5.7}$$

where

$$\lim_{\delta \to 0} \epsilon(n, \delta) = 0, \tag{14.5.8}$$

n being fixed. Therefore, denoting

$$||u_n(f) - y|| := \omega(n), \quad \lim_{n \to \infty} \omega(n) = 0, \tag{14.5.9}$$

we rewrite inequality (14.5.4) as

$$||u_n(f_\delta) - y|| \leq \epsilon(n, \delta) + \omega(n). \tag{14.5.10}$$

Minimizing the right-hand side of (14.5.10) with respect to n for a fixed small $\delta > 0$, one finds $n = n(\delta)$, and

$$\lim_{\delta \to 0} [\epsilon(n(\delta), \delta) + \omega(n(\delta))] = 0, \quad \lim_{\delta \to 0} n(\delta) = \infty. \tag{14.5.11}$$

One can also find $n(\delta)$ by solving the equation

$$\epsilon(n, \delta) = \omega(n) \tag{14.5.12}$$

for n for a fixed small $\delta > 0$. For a fixed $\delta > 0$ the quantity $\epsilon(n, \delta)$ grows as $n \to \infty$, so that equation (14.5.12) has a solution $n(\delta)$ with the properties (14.5.11). This was proved in Section 2.4 for linear operators.

For nonlinear operators one may choose an arbitrary small number $\eta > 0$, find $n = n(\eta)$, such that

$$\omega(n(\eta)) < \frac{\eta}{2},$$

then for a fixed $n(\eta)$ find $\delta = \delta(\eta)$ so small that

$$\epsilon(n(\eta), \delta(\eta)) < \frac{\eta}{2}.$$

Then

$$\epsilon(n(\eta), \delta(\eta)) + \omega(n(\eta)) < \eta.$$

The functions $n = n(\eta)$, $\delta = \delta(\eta)$ one can consider as a parametric representation of the dependence $n = n(\delta)$. The function $\omega(n)$ can be chosen without loss of generality monotonically decaying to zero as $n \to \infty$. Thus the equation $\omega(n) = \frac{\eta}{2}$ determines a unique monotone function $n = n(\eta)$, so that for a given n one can find a unique $\eta(n)$ such that $n(\eta(n)) = n$.

The function $\epsilon(n, \delta) = \delta p(n)$, where one may assume $\lim_{n\to\infty} p(n) = \infty$ and $p(n)$ grows monotonically when n grows.

Indeed

$$\epsilon(n, \delta) = ||u_n(f_\delta) - u_n(f)|| \leq \sup_{||\xi|| \in B(f,\delta)} ||u_n'(\xi)||\delta := p(n)\delta,$$

where $u_n'(\xi)$ is the Fréchet derivative of the operator $f \mapsto u_n(f)$ for fixed n. Since $p(n)$ is an upper bound on the norm of this operator, one may assume that $p(n)$ grows monotonically with n. One may minimize the right-hand side of (14.5.10) with respect to n for a fixed δ analytically if $\omega(n)$ and $p(n)$ are given analytically.

Chapter 15

Numerical problems arising in applications

15.1 Stable numerical differentiation

Let $f \in C^{\mu}(a, b)$, $\mu > 0$ be the space of μ times differentiable functions. If $0 < \mu < 1$, then the norm in $C^{\mu}(a, b)$ is defined as follows:

$$||f||_{C^{\mu}(a,b)} = \sup_{x \in (a,b)} |f(x)| + \sup_{x,y \in (a,b),\ x \neq y} \frac{|f(x) - f(y)|}{|x - y|^{\mu}}.$$

If $m < \mu < m + 1$, then

$$||f||_{C^{\mu}(a,b)} = \sup_{x \in (a,b)} \sum_{j=0}^{m} |f^{(j)}(x)| + \sup_{x,y \in (a,b),\ x \neq y} \frac{|f^{(m)}(x) - f^{(m)}(y)|}{|x - y|^{\mu - m}}.$$

If $\mu = m$, where $m \geq 0$ is an integer, then

$$||f||_{C^{m}(a,b)} = \sup_{x \in (a,b)} \sum_{j=0}^{m} |f^{(j)}(x)|.$$

Suppose that $\mu \geq 1$ and f_δ is given in place of f,

$$||f_\delta - f|| \leq \delta.$$

We are going to discuss the problem of stable numerical differentiation of f given the noisy data f_δ. By a stable approximation of f' one usually means an expression $R_\delta f_\delta$ such that

$$\lim_{\delta \to 0} ||R_\delta(f_\delta) - f'|| = 0, \tag{15.1.1}$$

where the norm is C^0-norm, i.e., the usual sup-norm. Thus

$$||R_\delta(f_\delta) - f'|| := \eta(\delta) \to 0 \quad \text{as} \quad \delta \to 0, \tag{15.1.2}$$

where R_δ is some, not necessarily linear, operator acting on the noisy data f_δ. Since the data are $\{\delta, f_\delta\}$ and the exact data f are unknown, we think that it is natural to change the standard definition of the regularizer and to define a regularizer R_δ by the relation:

$$\sup_{f \in B(f_\delta,\delta) \cap K} ||R_\delta(f_\delta) - f'|| := \eta(\delta) \to 0 \quad \text{as} \quad \delta \to 0, \tag{15.1.3}$$

where K is a compactum to which the exact data belong, and

$$B(f_\delta, \delta) := \{f : ||f - f_\delta|| \le \delta\}.$$

The above definition is the general Definition 2.1.3, applied to the problem of stable numerical differentiation.

In many applications K is defined as

$$K = K := \{f : ||f||_\mu \le M_\mu\}, \quad \mu > 1, \tag{15.1.4}$$

where $||\cdot||_\mu$ in (15.1.4) denotes $C^\mu-$norm.

We will prove that

If $\mu \le 1$, then there does not exist R_δ, linear or nonlinear, such that (15.1.3) holds. If $\mu > 1$, then a linear operator R_δ exists such that (15.1.3) holds.

Moreover, R_δ will be given explicitly, analytically, and $\eta(\delta)$ will be specified.

If $\mu = 2$, we will prove that R_δ that we construct is the best possible operator among all linear and nonlinear operators in the following sense:

$$\sup_{f \in K_{2,\delta}} ||R_\delta f_\delta - f'|| = \inf_{T \in N} \sup_{f \in K_{2,\delta}} ||T f_\delta - f'||, \tag{15.1.5}$$

where

$$K_{2,\delta} := K_2 \cap B(f_\delta, \delta),$$

and N is the set of all operators from $L^\infty(a,b)$ into $L^\infty(a,b)$.

Stable numerical differentiation is of practical interest in many applications. For example, if one measures the distance $s(t)$, traveled by a particle by the time t, and wants to find the velocity of this particle, one has to estimate $s'(t)$ stably given $s_\delta(t)$, $||s_\delta(t) - s(t)|| \le \delta$. The number of examples can be easily increased.

The first question to ask is:
When is it possible, in principle, to find R_δ satisfying (15.1.3)?
The second question we split into two questions:
How does one construct R_δ?
What is the error estimate in formula (15.1.3)?
Let us answer these questions.

Theorem 15.1.1. *If $\mu \leq 1$, then there is no operator R_δ such that (15.1.3) holds.*

Proof. Without loss of generality, let $(a, b) = (0, 1)$, and let

$$f_1(x) = -\frac{Mx(x-2h)}{2}, \quad 0 \leq x \leq 2h, \quad f_2(x) = -f_2(x). \tag{15.1.6}$$

Let us extend f_1 from $[0, 2h]$ to $[2h, 1]$ by zero and denote again by f_1 the extended function. This $f_1 \in W^{1,\infty}(0,1)$, where $W^{l,p}$ is the Sobolev space. We have

$$||f_k|| := \sup_{x\in(0,1)} |f_k(x)| = \frac{Mh^2}{2}, \qquad k = 1, 2. \tag{15.1.7}$$

Let

$$h = \sqrt{\frac{2\delta}{M}}. \tag{15.1.8}$$

Then

$$\frac{Mh^2}{2} = \delta.$$

For $f_\delta = 0$ we have

$$||f_k - f_\delta|| = ||f_k|| = \delta. \tag{15.1.9}$$

Denote

$$T(0)|_{x=0} := b, \tag{15.1.10}$$

where $T = T_\delta$ is some operator,

$$T : L^\infty(0,1) \to L^\infty(0,1).$$

Let K_1 be given by (15.1.4) with a constant M_1. We have

$$f'(0) = Mh.$$

Thus

$$\begin{aligned}\gamma &:= \inf_T \sup_{f\in K_{1,\delta}} ||T(f_\delta)-f'|| \geq \inf_b \max\{|b-f_1'(0)|, |b+f_1'(0)|\}\\ &= \inf_b \max\{|b-Mh|, |b+Mh|\}\\ &= Mh=\sqrt{2\delta M} := \epsilon(\delta). \qquad (15.1.11)\end{aligned}$$

If the constant M_1 in the definition of K_1 is fixed, then

$$\sup_{x\in(0,1)} |f_k'(x)| = \sup_{x\in(0,2h)} M|h-x| = Mh = \sqrt{2\delta M} \leq M_1. \qquad (15.1.12)$$

Since M in (15.1.6) is arbitrary, we can choose $M=M(\delta)$ so that $\sqrt{2\delta M} = M_1$, i.e. $M=\frac{M_1^2}{2\delta}$. In this case $\gamma \geq M_1 > 0$, so that relation (15.1.3) does not hold no matter what the choice of R_δ is. If $\mu < 1$, then for no choice of R_δ the relation (15.1.3) holds because f' does not exist in $L^\infty(0,1)$.

Theorem 15.1.1 is proved. □

An argument similar to the above leads to a similar conclusion if the derivatives are understood in L^2 sense, $||\cdot||$ is the norm in the Sobolev space $H^\mu = H^\mu(0,1)$, i.e. the space of real-valued functions with the norm defined for an integer $\mu \geq 0$ by the formula

$$||u||_\mu = \left(\sum_{j=0}^{\mu} ||u^{(j)}||_0^2\right)^{\frac{1}{2}},$$

where

$$||u||_0^2 := \int_0^1 |u|^2 dx,$$

and for arbitrary values $\mu > 0$ the norm is defined by interpolation, see, for instance [BL].

The main differences between C^μ and H^μ cases consists in the calculation of the norms $||f_1||_0$ and $||f_1'||_0$:

$$||f_1||_{L^2(0,1)} := ||f_1||_0 = c_0 M h^{\frac{5}{2}}, \quad ||f_1'|| = c_1 M h^{\frac{3}{2}}, \qquad (15.1.13)$$

where $c_0, c_1 > 0$ are constants.

Indeed

$$\begin{aligned}||f_1||_0^2 &= \int_0^{2h} \frac{M^2}{4} x^2 (x-2h)^2 dx = \frac{M^2h^5}{4}(2h)^2\left(\frac{1}{5}-\frac{1}{4}+\frac{1}{3}\right)\\ &= M^2h^5\frac{34}{15} := c_0^2 M^2 h^5,\end{aligned}$$

$$||f_1'||_0^2 = \int_0^{2h} M^2(x-h)^2dx = \frac{2}{3}M^2h^3 := c_1^2M^2h^3.$$

Choose h from the condition (15.1.9):

$$c_0Mh^{\frac{5}{2}} = \delta, \quad h = \left(\frac{\delta}{c_0M}\right)^{\frac{2}{5}}. \tag{15.1.14}$$

Let

$$f_\delta = 0, \quad K_{1,\delta}^{(2)} := \{f : ||f||_0 + ||f'||_0 \le M_1\}.$$

Then

$$\begin{aligned} \gamma_2 &:= \inf_T \sup_{f\in K_{1,\delta}^{(2)}} ||T(f_\delta) - f'|| \\ &\ge \inf_T \max\{||T(0) - f_1'||_0, ||T(0) - f_1'||_0\}. \end{aligned} \tag{15.1.15}$$

Denote

$$\varphi := T(0) \in L^2(0,1),$$

and set

$$\varphi = cf_1' + \psi, \quad (\psi, f_1') = 0, \quad c = \text{const}. \tag{15.1.16}$$

Then (15.1.15) implies

$$\begin{aligned} \gamma_2 &\ge \inf_{c\in\mathbb{R},\ \psi\perp f_1'} \max\{\sqrt{|1-c|^2||f_1'||_0^2 + ||\psi||_0^2}, \sqrt{|1+c|^2||f_1'||_0^2 + ||\psi||_0^2}\} \\ &= \inf_{c\in\mathbb{R}} \max\{|1-c|, |1+c|\}||f_1'||_0 = ||f_1'||_0 = c_1Mh^{\frac{3}{2}}. \end{aligned} \tag{15.1.17}$$

From (15.1.14) and (15.1.17) we get

$$\gamma_2 \ge \frac{c_1}{c_0^{\frac{3}{5}}} M^{\frac{2}{5}} \delta^{\frac{3}{5}}. \tag{15.1.18}$$

We choose h by formula (15.1.14), and then condition (15.1.9) holds. We then choose M such that

$$c_1Mh^{\frac{3}{2}} = M_1. \tag{15.1.19}$$

Then (15.1.17) yields

$$\gamma_2 \ge M_1 > 0. \tag{15.1.20}$$

Thus we have an analog of Theorem 15.1.1 for L^2- norm.

Theorem 15.1.2. *If $\mu \le 1$, then there is no operator R_δ such that (15.1.3) holds with $||\cdot|| = ||\cdot||_{L^2(0,1)}$.*

Proof. We have proved this theorem for $\mu = 1$. If $\mu < 1$, then f' does not exist in $L^2(0,1)$, so (15.1.3) does not hold with $||\cdot|| = ||\cdot||_{L^2(0,1)}$.

Let us now prove that if $\mu > 1$, then relation (15.1.3) holds. We obtain an estimate for $\eta_s(\delta)$. Define

$$R_\delta f_\delta := \begin{cases} \frac{f_\delta(x+h(\delta))-f_\delta(x)}{h(\delta)}, & 0 \le x \le h(\delta), \\ \frac{f_\delta(x+h(\delta))-f_\delta(x-h(\delta))}{2h(\delta)}, & h(\delta) \le x \le 1-h(\delta), \\ \frac{f_\delta(x)-f_\delta(x-h(\delta))}{h(\delta)}, & 1-h(\delta) \le x \le 1, \end{cases} \tag{15.1.21}$$

where $h(\delta) > 0$ will be specified below. Our argument is valid for L^p norms with any $p \in [1, \infty]$. We have

$$||R_\delta f_\delta - f'|| \le ||R_\delta f_\delta - R_\delta f|| + ||R_\delta f - f'|| \le ||R_\delta||\delta + ||R_\delta f - f'||. \tag{15.1.22}$$

For simplicity let us assume that f is periodic, $f(x+1) = f(x)$, and, therefore, f is defined on all of $\mathbb{R}$. Then we can use the middle line in (15.1.21) as the definition of $R_\delta f_\delta$, get

$$||R_\delta|| \le \frac{1}{h(\delta)}, \tag{15.1.23}$$

and

$$||R_\delta(f_\delta) - R_\delta(f)|| \le \frac{\delta}{h(\delta)}. \tag{15.1.24}$$

Let us estimate the last term in (15.1.22). Let $h(\delta) := h$, and assume $f \in C^2$. Then, for the L^∞-norm we have:

$$\begin{aligned} &||\tfrac{f(x+h)-f(x-h)}{2h} - f'|| \\ &= ||\tfrac{f(x)+hf'(x)+\frac{h^2}{2}f''(\xi_+)-f(x)+hf'(x)-\frac{h^2}{2}f''(\xi_-)}{2h} - f'|| \\ &\le \tfrac{M_2 h}{2}, \end{aligned} \tag{15.1.25}$$

where $\xi_\pm$ are the points in the remainder of the Taylor formula, and

$$M_m = \sup_{x\in(0,1)} |f^{(m)}(x)|.$$

Thus (15.1.22), (15.1.24) and (15.1.25) yield:

$$||R_\delta f_\delta - f'|| \le \frac{\delta}{h} + \frac{M_2 h}{2} := \eta(\delta, h). \tag{15.1.26}$$

Minimizing the right-hand side of (15.1.26) with respect to h, we get

$$h = \sqrt{\frac{2\delta}{M_2}}, \qquad \eta_s(\delta) := \eta(\delta, h(\delta)) = \sqrt{2M_2\delta}. \tag{15.1.27}$$

These formulas are obtained under the assumption $f \in C^2$.

If $f \in C^\mu$, $\quad 1 < \mu < 2$, then the estimate, analogous to (15.1.25), is:

$$\begin{aligned} \left\|\frac{f(x+h) - f(x-h)}{2h} - f'\right\| &= \left\|\frac{1}{2h}\int_{x-h}^{x+h} [f'(t) - f'(x)]dt\right\| \\ &\leq \frac{1}{2h}\left\|\int_{x-h}^{x+h} M_{\mu-1}|t-x|^{\mu-1}dt\right\| \\ &= \frac{2M_{\mu-1}}{2h}\left\|\int_0^h s^{\mu-1}ds\right\| \\ &= \frac{M_{\mu-1}h^{\mu-1}}{\mu}. \end{aligned} \tag{15.1.28}$$

Therefore estimates similar to (15.1.26) and (15.1.27) take the form:

$$\|R_\delta f_\delta - f'\| \leq \frac{\delta}{h} + \frac{M_{\mu-1}h^{\mu-1}}{\mu} = \eta(\delta, h), \tag{15.1.29}$$

$$\begin{aligned} h = h(\delta) = c_\mu \delta^{\frac{1}{\mu}}, \quad c_\mu &:= \left[\frac{\mu}{(\mu-1)M_{\mu-1}}\right]^{\frac{1}{\mu}}; \\ \eta(\delta) = \eta(\delta, h(\delta)) &= C_\mu \delta^{\frac{\mu-1}{\mu}}, \end{aligned} \tag{15.1.30}$$

where

$$C_\mu = \frac{1}{\mu} + \frac{M_{\mu-1}}{\mu} c_\mu^{\mu-1}. \tag{15.1.31}$$

□

We have proved the following result.

Theorem 15.1.3. *If $\mu \in (1,2)$, then R_δ, defined in (15.1.21), satisfies estimate (15.1.3) if $h = h(\delta)$ is given in (15.1.30), and then the error $\eta(\delta)$ is given also in (15.1.30).*

If $\mu > 2$, then one can define

$$R_h^{(Q)} := \frac{1}{h}\sum_{k=-Q}^{Q} A_k^{(Q)} f\left(x + \frac{kh}{Q}\right). \tag{15.1.32}$$

Suppose

$$\sup_x |f^{(m)}(x)| \le M_m, \tag{15.1.33}$$

where $m = 2q$ or $m = 2q+1$, and $q \ge 0$ is an integer. Let us take $Q = q$ in (15.1.32) and choose the coefficients $A_j^{(Q)}$ so that the difference $R_h^{(Q)} - f'$ has the highest order of smallness as $h \to 0$. Then these coefficients solve the following linear algebraic system:

$$\sum_{k=-Q}^{Q} \frac{1}{j!}\left(\frac{k}{Q}\right)^j A_k^{(Q)} = \delta_{1j}, \quad 0 \le j \le 2Q, \quad \delta_{kj} = \begin{cases} 1, & k = j, \\ 0, & k \ne j. \end{cases} \tag{15.1.34}$$

We set $A_0^{(Q)} = 0$. Then system (15.1.34) is a linear algebraic system for $2Q$ unknown coefficients $A_k^{(Q)}$, $k = \pm 1, \cdots, \pm Q$, with Vandermonde matrix whose determinant does not vanish. Thus, all the coefficients

$$A_k^{(Q)}, \quad 1 \le |k| \le Q,$$

are uniquely determined from the system (15.1.34). For example:

$$A_0^{(1)} = 0, \quad A_{\pm 1}^{(1)} = \pm\tfrac{1}{2}$$

$$A_0^{(2)} = 0, \quad A_{\pm 1}^{(2)} = \pm\tfrac{4}{3}, \quad A_{\pm 2}^{(2)} = \pm\tfrac{1}{6}$$

$$A_0^{(3)} = 0, \quad A_{\pm 1}^{(3)} = \pm\tfrac{9}{4}, \quad A_{\pm 2}^{(3)} = \pm\tfrac{9}{20}, \quad A_{\pm 3}^{(3)} = \pm\tfrac{1}{20},$$

etc.

Let $m = 2q+1 > 1, \quad q = Q$. We have

$$|R_h^{(Q)} f - f'| \le 2N_{2Q+1} h^{2Q}, \quad ||R_h^{(Q)}||_{L^\infty} = \frac{c(Q)}{h}, \tag{15.1.35}$$

where

$$c(Q) = \sum_{k=-Q}^{Q} |A_k^{(Q)}|.$$

Therefore an inequality analogous to (15.1.26) with L^∞-norm will be of the form

$$||R_h^{(Q)} f_\delta - f'|| \le \frac{c(Q)\delta}{h} + 2M_m h^{m-1} := \eta(\delta, h). \tag{15.1.36}$$

Minimizing the right-hand side of (15.1.36) with respect to h, we get:

$$h = h(\delta) = \left[\frac{c(Q)\delta}{2M_m(m-1)}\right]^{\frac{1}{m}} \delta^{\frac{1}{m}}, \tag{15.1.37}$$

and

$$\eta(\delta) = O\left(\delta^{1-\frac{1}{m}}\right) \quad \text{as} \quad \delta \to 0.$$

We have proved the following result.

Theorem 15.1.4. *If $f \in C^m$, $m > 2$, then the operator $R_\delta := R^{(Q)}_{h(\delta)}$, where $Q = \frac{m-1}{2}$ if m is odd, $Q = \frac{m}{2}$ if m is even, and $h(\delta)$ is given in (15.1.37), yields a stable approximation $R_\delta f_\delta$ of f' with the error $\eta(\delta) = O(\delta^{1-\frac{1}{m}})$ as $\delta \to 0$, and estimate (15.1.3) holds. A similar result holds for L^p-norm, $p \in [1, \infty]$.*

Finally we state the following result, which follows from (15.1.11) and (15.1.30) with $\mu = 2$:

$$||R_\delta f_\delta - f'|| \leq \sqrt{2M_2\delta}, \quad h(\delta) = \sqrt{\frac{2\delta}{M_2}}. \tag{15.1.38}$$

Let $\mu = 2$ and the norm be L^∞-norm. Then, among all operators T, linear and nonlinear, acting from the space of L^∞ periodic functions with period 1 into itself, the operator

$$Tf_\delta = R_\delta f_\delta := \frac{f_\delta(x + h(\delta)) - f_\delta(x - h(\delta))}{2h(\delta)}$$

with $h(\delta) = \sqrt{\frac{2\delta}{M_2}}$ gives the best possible estimate of f', given the data f_δ, $||f_\delta - f|| \leq \delta$, on the class of all f such that $\sup_x |f''(x)| \leq M_2$.

15.2 Stable differentiation of piecewise smooth functions

Let f be a piecewise-$C^2([0, 1])$ function,

$$0 < x_1 < x_2 < \cdots < x_J, \quad 1 \leq j \leq J,$$

be the discontinuity points of f. We do not assume their locations x_j and their number J known a priori. We assume that the limits $f(x_j \pm 0)$ exist, and

$$\sup_{x \neq x_j, 1 \leq j \leq J} |f^{(m)}(x)| \leq M_m, \quad m = 0, 1, 2. \tag{15.2.1}$$

Assume that f_δ is given,

$$\|f - f_\delta\| := \sup_{x \neq x_j, 1 \leq j \leq J} |f - f_\delta| \leq \delta,$$

where $f_\delta \in L^\infty(0,1)$ are the noisy data.

The problem is:

Given $\{f_\delta, \delta\}$, where $\delta \in (0, \delta_0)$ and $\delta_0 > 0$ is a small number, estimate stably f', find the locations of discontinuity points x_j of f and their number J, and estimate the jumps

$$p_j := f(x_j + 0) - f(x_j - 0)$$

of f across x_j, $1 \leq j \leq J$.

A stable estimate $R_\delta f_\delta$ of f' is an estimate satisfying the relation

$$\lim_{\delta \to 0} ||R_\delta f_\delta - f'|| = 0.$$

There is a large literature on stable differentiation of noisy smooth functions, but the problem stated above was not solved for piecewise-smooth functions by the method given below. A statistical estimation of the location of discontinuity points from noisy discrete data is given in [KR2]. In [R25], [R21], [KR1], various approaches to finding discontinuities of functions from the measured values of these functions are developed.

The following formula (see Section 15.1) was proposed originally (in 1968, see [R4]) for stable estimation of $f'(x)$, assuming $f \in C^2([0,1])$, $M_2 \neq 0$, and given noisy data f_δ:

$$\begin{aligned} R_\delta f_\delta &:= \frac{f_\delta(x + h(\delta)) - f_\delta(x - h(\delta))}{2h(\delta)}, \\ h(\delta) &:= \left(\frac{2\delta}{M_2}\right)^{\frac{1}{2}}, \quad h(\delta) \leq x \leq 1 - h(\delta), \end{aligned} \tag{15.2.2}$$

and

$$\|R_\delta f_\delta - f'\| \leq \sqrt{2M_2\delta} := \varepsilon(\delta), \tag{15.2.3}$$

where the norm in (15.2.3) is $L^\infty(0,1)-$norm. Numerical efficiency and stability of the stable differentiation method proposed in [R4] has been demonstrated in [R30]. Moreover, (cf [R13], p. 345, and Section 15.1),

$$\inf_T \sup_{f \in K(M_2, \delta)} \|Tf_\delta - f'\| \geq \varepsilon(\delta), \tag{15.2.4}$$

where $T : L^\infty(0,1) \to L^\infty(0,1)$ runs through the set of all bounded operators,

$$K(M_2, \delta) := \{f : \|f''\| \leq M_2,\ \|f - f_\delta\| \leq \delta\}.$$

Therefore estimate (15.2.2) is the best possible estimate of f', given noisy data f_δ, and assuming $f \in K(M_2, \delta)$.

In [R44] this result was generalized to the case

$$f \in K(M_a, \delta), \quad \|f^{(a)}\| \leq M_a, \quad 1 < a \leq 2,$$

where

$$\|f^{(a)}\| := \|f\| + \|f'\| + \sup_{x,x'} \frac{|f'(x) - f'(x')|}{|x - x'|^{a-1}}, \quad 1 < a \leq 2,$$

and $f^{(a)}$ is the fractional-order derivative of f.

The aim of this Section is to extend the above results to the case of piecewise-smooth functions.

Theorem 15.2.1. *Formula* (15.2.2) *gives stable estimate of* f' *on the set* $S_\delta := [h(\delta), 1 - h(\delta)] \setminus \bigcup_{j=1}^{J} (x_j - h(\delta), x_j + h(\delta))$, *and* (15.2.3) *holds with the norm* $\|\cdot\|$ *taken on the set* S_δ. *Assuming* $M_2 > 0$ *and computing the quantities*

$$f_j := \frac{f_\delta(jh + h) - f_\delta(jh - h)}{2h},$$

where

$$h := h(\delta) := \left(\frac{2\delta}{M_2}\right)^{\frac{1}{2}}, \quad 1 \leq j < [\frac{1}{h}],$$

for sufficiently small δ, *one finds the location of discontinuity points of* f *with accuracy* $2h$, *and their number* J. *Here* $[\frac{1}{h}]$ *is the integer smaller than* $\frac{1}{h}$ *and closest to* $\frac{1}{h}$. *The discontinuity points of* f *are located on the intervals* $(jh - h, jh + h)$ *such that* $|f_j| \gg 1$ *for sufficiently small* δ, *where* $\varepsilon(\delta)$ *is defined in (15.2.3) . The size* p_j *of the jump of* f *across the discontinuity point* x_j *is estimated by the formula*

$$p_j \approx f_\delta(jh + h) - f_\delta(jh - h),$$

and the error of this estimate is $O(\sqrt{\delta})$.

Let us assume that $\min_j |p_j| := p \gg h(\delta)$, where $\gg$ means "much greater than". Then x_j is located on the j-th interval $[jh - h, jh + h]$, $h := h(\delta)$, such that

$$|f_j| := \left| \frac{f_\delta(jh+h) - f_\delta(jh-h)}{2h} \right| \gg 1, \tag{15.2.5}$$

so that x_j is localized with the accuracy $2h(\delta)$. More precisely,

$$|f_j| \geq \frac{|f(jh+h) - f(jh-h)|}{2h} - \frac{\delta}{h},$$

and

$$\frac{\delta}{h} = 0.5\varepsilon(\delta),$$

where $\varepsilon(\delta)$ is defined in (15.2.3) . One has

$$\begin{aligned} |f(jh+h) - f(jh-h)| &\geq |p_j| - |f(jh+h) - f(x_j+0)| \\ &\quad -|f(jh-h) - f(x_j-0)|, \\ &\geq |p_j| - 2M_1 h. \end{aligned}$$

Thus,

$$|f_j| \geq \frac{|p_j|}{2h} - M_1 - 0.5\varepsilon(\delta) = c_1 \frac{|p_j|}{\sqrt{\delta}} - c_2 \gg 1,$$

where

$$c_1 := \frac{\sqrt{M_2}}{2\sqrt{2}}, \quad \text{and} \quad c_2 := M_1 + 0.5\varepsilon(\delta).$$

The jump p_j is estimated by the formula:

$$p_j \approx [f_\delta(jh+h) - f_\delta(jh-h)], \tag{15.2.6}$$

and the error estimate of this formula can be given:

$$\begin{aligned} |p_j - [f_\delta(jh+h) - f_\delta(jh-h)]| &\leq 2\delta + 2M_1 h \\ &= 2\delta + 2M_1\sqrt{\tfrac{2\delta}{M_2}} = O(\sqrt{\delta}). \end{aligned} \tag{15.2.7}$$

Thus, the error of the calculation of p_j by the formula

$$p_j \approx f_\delta(jh+h) - f_\delta(jh-h)$$

is $O(\delta^{\frac{1}{2}})$ as $\delta \to 0$.

Proof of Theorem 15.2.1. If $x \in S_\delta$, then, using Taylor's formula, one gets:

$$|(R_\delta f_\delta)(x) - f'(x)| \leq \frac{\delta}{h} + \frac{M_2 h}{2}. \tag{15.2.8}$$

Here we assume that $M_2 > 0$ and the interval $(x - h(\delta), x + h(\delta)) \subset S_\delta$, i.e., this interval does not contain discontinuity points of f. If for all sufficiently small h, not necessarily for $h = h(\delta)$, inequality (15.2.8) fails, i.e., if

$$|(R_\delta f_\delta)(x) - f'(x)| > \frac{\delta}{h} + \frac{M_2 h}{2}$$

for all sufficiently small $h > 0$, then the interval $(x - h, x + h)$ contains a point $x_j \notin S_\delta$, i.e., a point of discontinuity of f or f'. This observation can be used for locating the position of an isolated discontinuity point x_j of f with any desired accuracy provided that the size $|p_j|$ of the jump of f across x_j is greater than $k\delta$, where $k > 2$ is a constant, $|p_j| > k\delta$, and that h can be taken as small as desirable.

Indeed, if $x_j \in (x - h, x + h)$, then we have

$$|p_j| - 2\delta - 2hM_1 \leq |f_\delta(x + h) - f_\delta(x - h)| \leq |p_j| + 2hM_1 + 2\delta.$$

The above estimate follows from the relation

$$\begin{aligned} &|f_\delta(x + h) - f_\delta(x - h)| \\ &= |f(x + h) - f(x_j + 0) + p_j + f(x_j - 0) - f(x - h) \pm 2\delta| \\ &= |p_j \pm (2hM_1 + 2\delta)|. \end{aligned}$$

Here $|p \pm b|$, where $b > 0$, denotes a quantity such that

$$|p| - b \leq |p \pm b| \leq |p| + b.$$

Thus, if h is sufficiently small and $|p_j| > k\delta$, where $k > 2$, then the inequality

$$(k - 2)\delta - 2hM_1 < |f_\delta(x + h) - f_\delta(x - h)|$$

can be checked, and therefore the inclusion $x_j \in (x - h, x + h)$ can be checked. Since $h > 0$ is arbitrarily small in this argument, it follows that the location of the discontinuity point x_j of f, at which $|p_j| > k\delta$ with $k > 2$, can be established with arbitrary accuracy.

A discussion of the case when a discontinuity point x_j belongs to the interval $(x - h(\delta), x + h(\delta))$ will be given below.

Minimizing the right-hand side of (15.2.8) with respect to h yields formula (15.2.2) for the minimizer $h = h(\delta)$ defined in (15.2.2), and estimate (15.2.3) for the minimum of the right-hand side of (15.2.8).

If $|p| \gg h(\delta)$, and (15.2.5) holds, then the discontinuity points are located with the accuracy $2h(\delta)$, as we prove now by an argument very similar to the one given above.

Consider the case when a discontinuity point x_j of f belongs to the interval $(jh - h, jh + h)$, where $h = h(\delta)$. Then estimate (15.2.6) can be obtained as follows. For $jh - h \leq x_j \leq jh + h$, one has

$$\begin{aligned} |f(x_j + 0) - f(x_j - 0) - f_\delta(jh + h) + f_\delta(jh - h)| \\ \leq 2\delta + |f(x_j + 0) - f(jh + h)| \\ +|f(x_j - 0) - f(jh - h)| \leq 2\delta + 2hM_1, \quad h = h(\delta). \end{aligned}$$

This yields formulas (15.2.6) and (15.2.7). Computing the quantities f_j for $1 \leq j < [\frac{1}{h}]$, and finding the intervals on which (15.2.5) holds for sufficiently small δ, one finds the location of discontinuity points of f with accuracy $2h$, and the number J of these points. For a small fixed $\delta > 0$ the above method allows one to recover the discontinuity points of f at which

$$|f_j| \geq \frac{|p_j|}{2h} - \frac{\delta}{h} - M_1 \gg 1.$$

This is the inequality (15.2.5). If $h = h(\delta)$, then

$$\frac{\delta}{h} = 0.5\varepsilon(\delta) = O(\sqrt{\delta}),$$

and

$$|2hf_j - p_j| = O(\sqrt{\delta}) \quad \text{as} \quad \delta \to 0,$$

provided that $M_2 > 0$. Theorem 15.2.1 is proved. □

Remark 15.2.1. Similar results can be derived if

$$\|f^{(\mu)}\|_{L^\infty(S_\delta)} := \|f^{(\mu)}\|_{S_\delta} \leq M_\mu, \; 1 < \mu \leq 2.$$

In this case

$$h = h(\delta) = c_\mu \delta^{\frac{1}{\mu}},$$

where $c_\mu = \left[\frac{2}{M_\mu(\mu-1)}\right]^{\frac{1}{\mu}}$, $R_\delta f_\delta$ is defined in (15.2.2), and the error of the estimate is:

$$\|R_\delta f_\delta - f'\|_{S_\delta} \leq \mu M_\mu^{\frac{1}{\mu}} \left(\frac{2}{\mu - 1}\right)^{\frac{\mu-1}{\mu}} \delta^{\frac{\mu-1}{\mu}}.$$

The proof is similar to the given above. It is proved in Section 15.1 that for C^{μ}-functions given with noise it is possible to construct stable differentiation formulas if $\mu > 1$ and it is impossible to construct such formulas if $\mu \leq 1$. The obtained formulas are useful in applications. One can also use L^p-norm on S_δ in the estimate $\|f^{(\mu)}\|_{S_\delta} \leq M_\mu$.

Remark 15.2.2. The case when $M_2 = 0$ requires a discussion. In this case the last term on the right-hand side of formula (15.2.8) vanishes and the minimization with respect to h becomes meaningless: it requires that h be as large as possible, but one cannot take h arbitrarily large because estimate (15.2.8) is valid only on the interval $(x - h, x + h)$ which does not contain discontinuity points of f, and these points are unknown. If $M_2 = 0$, then f is a piecewise-linear function. The discontinuity points of a piecewise-linear function can be found if the sizes $|p_j|$ of the jumps of f across these points satisfy the inequality

$$|p_j| >> 2\delta + 2hM_1$$

for some choice of h. This follows from the estimate

$$|f_\delta(jh + h) - f_\delta(jh - h)| \geq |p_j| - 2hM_1 - 2\delta,$$

if the discontinuity point x_j lies on the interval $(jh-h, jh+h)$. For instance, if $h = \frac{\delta}{M_1}$, then $2\delta + 2M_1 h = 4\delta$. So, if $|p_j| >> 4\delta$, then the location of discontinuity points of f can be found in the case when $M_2 = 0$. The discontinuity points x_j of f are located on the intervals for which

$$|f_\delta(jh + h) - f_\delta(jh - h)| >> 4\delta,$$

where $h = \frac{\delta}{M_1}$.

The size $|p_j|$ of the jump of f across a discontinuity point x_j can be estimated by formula (15.2.6) with $h = \frac{\delta}{M_1}$, and one assumes that $x_j \in (jh-h, jh+h)$ is the only discontinuity point on this interval. The error of the formula (15.2.6) is estimated as in the proof of Theorem 15.2.1. This error is not more than $2\delta + 2M_1 h = 4\delta$ for the above choice of $h = \frac{\delta}{M_1}$.

One can estimate the derivative of f at the point of smoothness of f assuming $M_2 = 0$ provided that this derivative is not too small. If $M_2 = 0$, then $f = a_j x + b_j$ on every interval Δ_j between the discontinuity points x_j, where a_j and b_j are some constants. If $(jh - h, jh + h) \subset \Delta_j$, and

$$f_j := \frac{f_\delta(jh + h) - f_\delta(jh - h)}{2h},$$

then

$$|f_j - a_j| \leq \frac{\delta}{h}.$$

Choose $h = \frac{t\delta}{M_1}$, where $t > 0$ is a parameter, and $M_1 = \max_j |a_j|$. Then the relative error of the approximate formula

$$a_j \approx f_j$$

for the derivative $f' = a_j$ on Δ_j equals to

$$\frac{|f_j - a_j|}{|a_j|} \leq \frac{M_1}{t|a_j|}.$$

Thus, if, e.g., $|a_j| \geq \frac{M_1}{2}$ and $t = 20$, then the relative error of the above approximate formula is not more than 0.1.

Suppose now that $\xi \in (mh-h, mh+h)$, where $m > 0$ is an integer, and *ξ is a point at which f is continuous but $f'(\xi)$ does not exist.* Thus, the jump of f across ξ is zero, but ξ is not a point of smoothness of f. *How does one locate the point ξ?*

The algorithm we propose consists of the following. We assume that $M_2 > 0$ on S_δ. Calculate the numbers

$$f_j := \frac{f_\delta(jh+h) - f_\delta(jh-h)}{2h}$$

and

$$|f_{j+1} - f_j|, \quad j = 1, 2, \ldots, \qquad h = h(\delta) = \sqrt{\frac{2\delta}{M_2}}.$$

Inequality (15.2.3) implies

$$f_j - \varepsilon(\delta) \leq f'(jh) \leq f_j + \varepsilon(\delta),$$

where $\varepsilon(\delta)$ is defined in (15.2.3).

Therefore, if

$$|f_j| > \varepsilon(\delta),$$

then

$$\text{sign } f_j = \text{ sign } f'(jh).$$

One has:

$$J - \frac{2\delta}{h} \leq |f_{j+1} - f_j| \leq J + \frac{2\delta}{h},$$

where

$$\frac{\delta}{h} = 0.5\varepsilon(\delta)$$

and

$$J := \left|\frac{f(jh+2h) - f(jh) - f(jh+h) + f(jh-h)}{2h}\right|.$$

Using Taylor's formula, one derives the estimate:

$$0.5[J_1 - \varepsilon(\delta)] \leq J \leq 0.5[J_1 + \varepsilon(\delta)], \tag{15.2.9}$$

where

$$J_1 := |f'(jh+h) - f'(jh)|.$$

If the interval $(jh-h, jh+2h)$ belongs to S_δ, then

$$J_1 = |f'(jh+h) - f'(jh)| \leq M_2 h = \varepsilon(\delta).$$

In this case

$$J \leq \varepsilon(\delta),$$

so

$$|f_{j+1} - f_j| \leq 2\varepsilon(\delta) \quad \text{if} \quad (jh-h, jh+2h) \subset S_\delta. \tag{15.2.10}$$

Conclusion: If

$$|f_{j+1} - f_j| > 2\varepsilon(\delta),$$

then the interval $(jh-h, jh+2h)$ does not belong to S_δ, that is, there is a point $\xi \in (jh-h, jh+2h)$ at which the function f is not twice continuously differentiable with $|f''| \leq M_2$. Since we assume that either at a point ξ the function is twice differentiable, or at this point f' does not exist, it follows that if $|f_{j+1} - f_j| > 2\varepsilon(\delta)$, then there is a point $\xi \in (jh-h, jh+2h)$ at which f' does not exist.

If

$$f_j f_{j+1} < 0, \tag{15.2.11}$$

and

$$\min(|f_{j+1}|, |f_j|) > \varepsilon(\delta), \tag{15.2.12}$$

then (15.2.11) implies $f'(jh)f'(jh+h) < 0$, so the interval $(jh, jh+h)$ contains a critical point ξ of f, or a point ξ at which f' does not exist. To determine which one of these two cases holds, let us use the right inequality

(15.2.9). If ξ is a critical point of f and $\xi \in (jh, jh+h) \subset S_\delta$, then $J_1 \leq \varepsilon(\delta)$, and in this case the right inequality (15.2.9) yields $J \leq \varepsilon(\delta)$. Thus

$$|f_{j+1} - f_j| \leq 2\varepsilon(\delta). \tag{15.2.13}$$

Conclusion: If (15.2.11) - (15.2.13) hold, then ξ is a critical point. If (15.2.11) and (15.2.12) hold and

$$|f_{j+1} - f_j| > 2\varepsilon(\delta),$$

then ξ is a point of discontinuity of f'.

If ξ is a point of discontinuity of f', we would like to estimate the jump

$$P := |f'(\xi + 0) - f'(\xi - 0)|.$$

Using Taylor's formula one gets

$$f_{j+1} - f_j = \frac{P}{2} \pm 2.5\varepsilon(\delta). \tag{15.2.14}$$

The expression $A = B \pm b,\ b > 0$, means that $B - b \leq A \leq B + b$. Therefore,

$$P = 2(f_{j+1} - f_j) \pm 5\varepsilon(\delta). \tag{15.2.15}$$

We have proved the following theorem:

Theorem 15.2.2. *If $\xi \in (jh-h, jh+2h)$ is a point of continuity of f and $|f_{j+1} - f_j| > 2\varepsilon(\delta)$, then ξ is a point of discontinuity of f'. If (15.2.11) and (15.2.12) hold, and $|f_{j+1} - f_j| \leq 2\varepsilon(\delta)$, then ξ is a critical point of f. If (15.2.11) and (15.2.12) hold and $|f_{j+1} - f_j| > 2\varepsilon(\delta)$, then $\xi \in (jh, jh+h)$ is a point of discontinuity of f'. The jump P of f' across ξ is estimated by formula (15.2.15) .*

Let us give a method for finding *nonsmoothness points of piecewise-linear functions.*

Assume that f is a piecewise-linear function on the interval $[0,1]$ and $0 < x_1 < \cdots < x_J < 1$ is its nonsmoothness points, i.e, the discontinuity points of f or these of f'. Assume that f_δ is known at a grid mh, $m = 0, 1, 2, \ldots\ldots, M$,

$$h = \frac{1}{M}, \quad f_{\delta,m} = f_\delta(mh), \quad |f(mh) - f_{\delta,m}| \leq \delta \ \forall m,$$

where $f_m = f(mh)$. If mh is a discontinuity point, $mh = x_j$, then we define its value as $f(x_j - 0)$ or $f(x_j + 0)$, depending on which of these two numbers satisfy the inequalty $|f(mh) - f_{\delta,m}| \leq \delta$.

The problem is:

Given $f_{\delta,m}$ $\forall m$, *estimate the location of the discontinuity points* x_j, *their number* J, *find out which of these points are points of discontinuity of* f *and which are points of discontinuity of* f' *but points of continuity of* f, *and estimate the sizes of the jumps of* f

$$|p_j| = |f(x_j + 0) - f(x_j - 0)|$$

and the sizes of the jumps of f'

$$q_j = |f'(x_j + 0) - f'(x_j - 0)|$$

at the continuity points of f *which are discontinuity points of* f'.

Let us solve this problem. Consider the quantities

$$G_m := \frac{f_{\delta,m+1} - 2f_{\delta,m} + f_{\delta,m-1}}{2h^2} := g_m + w_m,$$

where

$$g_m := \frac{f_{m+1} - 2f_m + f_{m-1}}{2h^2},$$

and

$$w_m := \frac{f_{\delta,m+1} - f_{m+1} - 2(f_{\delta,m} - f_m) + f_{\delta,m-1} - f_m}{2h^2}.$$

We have

$$|w_m| \le \frac{4\delta}{2h^2} = \frac{2\delta}{h^2},$$

and

$$g_m = 0 \text{ if } x_j \notin (mh - h, mh + h) \quad \forall j.$$

Therefore, the following claim hold:

Claim:

If $\min_j |x_{j+1} - x_j| > 2h$ *and*

$$|G_m| > \frac{2\delta}{h^2}, \tag{15.2.16}$$

then the interval $(mh - h, mh + h)$ *must contain a discontinuity point of* f.

Condition (15.2.16) is sufficient for the interval $(mh - h, mh + h)$ to contain a discontinuity point of f, but is not a necessary condition: it may happen that the interval $(mh - h, mh + h)$ contains more than one discontinuity points (this is only possible if the assumption $\min_j |x_{j+1} - x_j| > 2h$ does not hold) without changing g_m or G_m, so that one cannot detect these points by the above method. We have proved the following result.

Theorem 15.2.3. *Condition (15.2.16) is a sufficient condition for the interval $(mh-h, mh+h)$ to contain a nonsmoothness point of f. If one knows a priori that $\min_j |x_{j+1}-x_j| > 2h$, then condition (15.2.16) is a necessary and sufficient condition for the interval $(mh-h, mh+h)$ to contain exactly one point of nonsmoothness of f.*

Let us estimate the size of the jump $|p_j| = |f(x_j+0) - f(x_j-0)|$. Let us assume that (15.2.16) holds, $x_{j+1} - x_j > 2h$, so there is only one discontinuity point x_j of f on the interval $(mh-h, mh+h)$, and assume that $x_j \in (mh-h, mh)$. The case when $x_j \in (mh, mh+h)$ is treated similarly. Let

$$f(x) = a_j x + b_j \quad \text{when} \quad mh < x < x_j,$$

and

$$f(x) = a_{j+1}x + b_{j+1} \quad \text{when} \quad x_j < x < (m+1)h,$$

where a_j, b_j are constants. One has

$$g_m = \frac{-(a_{j+1}-a_j)(mh-h) - (b_{j+1}-b_j)}{2h^2},$$

and

$$|p_j| = |(a_{j+1}-a_j)x_j + b_{j+1} - b_j|.$$

Thus

$$|g_m| = \left|\frac{-(a_{j+1}-a_j)x_j - (b_{j+1}-b_j) - (a_{j+1}-a_j)(mh-h-x_j)}{2h^2}\right|$$

$$= \frac{|p_j|}{2h^2} \pm \frac{|a_{j+1}-a_j||x_j-(mh-h)|}{2h^2},$$

where $A \pm B$ denotes a quantity such that $A - B \leq A \pm B \leq A + B$, $A, B > 0$.

Let

$$|a_{j+1} - a_j| = q_j.$$

Note that

$$|x_j - (mh-h)| \leq h \quad \text{if} \quad mh - h < x_j < mh.$$

Thus,

$$|G_m| = \frac{|p_j|}{2h^2} \pm \left(\frac{q_j h}{2h^2} + \frac{2\delta}{h^2}\right).$$

Therefore,

$$|G_m| = \frac{|p_j|}{2h^2}\left[1 \pm \left(\frac{q_j h}{|p_j|} + \frac{4\delta}{|p_j|}\right)\right],$$

provided that $|p_j| > 0$.

If

$$\frac{q_j h + 4\delta}{|p_j|} << 1 \text{ and } |p_j| > 0,$$

then

$$|p_j| \approx 2h^2 |G_m|.$$

If $p_j = 0$, then $x_j = \frac{b_j - b_{j+1}}{a_{j+1} - a_j}$, and

$$|G_m| = \frac{q_j(x_j - mh + h)}{2h^2} \pm \frac{2\delta}{h^2},$$

because $x_j > mh - h$ by the assumption. Thus,

$$q_j \approx \frac{2h^2 |G_m|}{x_j - mh + h},$$

provided that $p_j = 0$ and $\delta << q_j(x_j - mh + h)$.

If

$$\min_j |x_{j+1} - x_j| > 2h,$$

then the number J of the nonsmoothness points of f can be determined as the number of intervals on which (15.2.16) holds.

15.3 Simultaneous approximation of a function and its derivative by interpolation polynomials

In this Section we present a result from [R8]. We want to construct an interpolation polynomial which gives a stable approximation of a continuous function $f \in C(I)$, $I := [-1, 1]$, given noisy values $f_\delta(x_j)$, $|f_\delta(x_j) - f(x_j)| < \delta$, and approximation should have the property that its derivative gives a stable approximation of f'.

The idea is to use Lagrange interpolating polynomial $L_n(x)$ with nodes at the roots

$$x_j = x_{j,n+1} := \cos\left(\frac{2j-1}{2n+2}\pi\right), \quad 1 \le j \le n+1,$$

of the Tchebyshev polynomial $T_{n+1}(x), \quad x \in I$,

$$T_n(x) := 2^{-(n-1)} \cos(n \arccos x), \quad L_n = L_n(x, f) = \sum_{j=1}^{n+1} f(x_j) l_j(x), \tag{15.3.1}$$

where

$$l_j(x) := \frac{T_{n+1}(x)}{T'_{n+1}(x_j)(x - x_j)}. \tag{15.3.2}$$

Let

$$\lambda_n := \max_{x \in I} \sum_{j=1}^{n+1} |l_j(x)|, \quad \lambda'_n(x) := \sum_{j=1}^{n+1} |l'_j(x)|, \quad \lambda'_n := \max_{x \in I} \lambda'_n(x). \tag{15.3.3}$$

It is known (see [A]) that

$$\frac{\ln(n+1)}{8\sqrt{\pi}} < \lambda_n < 8 + \frac{4}{\pi} \ln(n+1). \tag{15.3.4}$$

Let

$$E_n(f) := \min_{P \in \mathcal{P}_n} \max_{x \in I} |f(x) - P(x)|,$$

where $\mathcal{P}_n$ is the set of all polynomials of degree $\leq n$. The polynomial of the best approximation exists and is unique for any $f \in C(I)$.

If $C^r = C^r(I)$ is the set of r times continuously differentiable functions on I with the norm

$$||f||_r = \max_{x \in I} \{|f(x)| + |f^{(r)}(x)|\}$$

and $f \in C^1$, then (see [T])

$$||f - L_n|| \leq (1 + \lambda_n) E_n(f), \tag{15.3.5}$$

where $|| \cdot || := || \cdot ||_0$.

Let us state our results.

Theorem 15.3.1. *Let $f \in C^1$. Then*

$$||f' - L'_n|| \leq (1 + \lambda'_n) E_n(f'), \tag{15.3.6}$$

$$|f'(x) - L'_n(x)| \leq \left(1 + \frac{n\lambda_n}{\sqrt{1 - x^2}}\right) E_{n-1}(f'), \quad \lambda'_n \leq n^2 \lambda_n, \tag{15.3.7}$$

$$\lambda'_n(x) \leq \frac{n\lambda_n}{\sqrt{1 - x^2}}. \tag{15.3.8}$$

Let

$$L_{n,\delta}(x) := \sum_{j=1}^{n+1} f_\delta(x_j) l_j(x).$$

Our second result is the following theorem:

Theorem 15.3.2. *Let $f \in C^r$, $r > 3$. Then there exists a function $n(\delta)$ such that*

$$||L_{n(\delta),\delta} - f|| = O(\delta |\ln \delta|), \quad \delta \to 0, \tag{15.3.9}$$

$$|\frac{d}{dx} L_{n(\delta),\delta}(x) - f'(x)| \le E_{n-1}(f')[1 + \lambda_n'] + \delta \lambda_n'(x), \tag{15.3.10}$$

$$||L_{n(\delta),\delta} - f|| \le (1 + \lambda_n) E_n(f) + \delta \lambda_n. \tag{15.3.11}$$

The following lemma will be used in the proofs.

Lemma 15.3.1. *Let $P_j \in \mathcal{P}_n$ and*

$$\sum_{j=1}^{m} |P_j(x)| \le M, \quad x \in I, \tag{15.3.12}$$

where $M = const$ and $m \ge 1$ is an arbitrary integer. Then

$$\sum_{j=1}^{m} |P_j'(x)| \le \frac{Mn}{\sqrt{1 - x^2}}, \tag{15.3.13}$$

$$\sum_{j=1}^{m} |P_j'(x)| \le M n^2. \tag{15.3.14}$$

This lemma is a generalization of the known for $m = 1$ inequalities of S. Bernstein.

Proof of Lemma 15.3.1. Let

$$x = \cos \theta, \quad P_j(x) = \mathcal{P}_j(\theta).$$

Then (see [T], p. 227):

$$\mathcal{P}_j'(\theta) = \frac{1}{4n} \sum_{k=1}^{2n} (-1)^{k+1} \mathcal{P}_j(\theta + \theta_k) \frac{1}{\sin^2(\frac{\theta_k}{2})}, \quad \theta_k = \frac{(2k-1)\pi}{2n}, \tag{15.3.15}$$

and

$$\frac{1}{4n}\sum_{k=1}^{2n}\frac{1}{\sin^2\left(\frac{\theta_k}{2}\right)} = n. \tag{15.3.16}$$

From (15.3.12), (15.3.15) and (15.3.16) we get

$$\sum_{j=1}^{m}|\mathcal{P}_j'(\theta)| \le \sum_{j=1}^{m}\left|\frac{1}{4n}\sum_{k=1}^{2n}(-1)^{k+1}\mathcal{P}_j(\theta+\theta_k)\frac{1}{\sin^2(\frac{\theta_k}{2})}\right| \tag{15.3.17}$$

$$\le \frac{1}{4n}\sum_{k=1}^{2n}(-1)^{k+1}\frac{1}{\sin^2(\frac{\theta_k}{2})}\sum_{j=1}^{m}|\mathcal{P}_j(\theta+\theta_k)| \le Mn. \tag{15.3.18}$$

Using (15.3.18), we get

$$\sum_{j=1}^{m}|P_j'(x)| = \sum_{j=1}^{m}|\mathcal{P}_j'(\theta)|\left|\frac{d\theta}{dx}\right| \le \frac{Mn}{\sqrt{1-x^2}}, \tag{15.3.19}$$

so (15.3.13) is proved.

Let us prove (15.3.14).

If $|x| \le \cos\left(\frac{\pi}{2n}\right)$, then, using the inequality

$$\sin x \ge \frac{2x}{\pi}, \quad 0 \le x \le \frac{\pi}{2},$$

we get:

$$\sqrt{1-x^2} \ge \sin\left(\frac{\pi}{2n}\right) > \frac{\pi}{2n}\frac{2}{\pi} = \frac{1}{n}, \tag{15.3.20}$$

so (15.3.14) follows from (15.3.19) if $|x| \le \cos\left(\frac{\pi}{2n}\right)$.

If

$$\cos\left(\frac{\pi}{2n}\right) \le |x| \le 1,$$

then we use the following known formula: (see [PS], problem VI. 71)

$$P_k'(x) = \frac{2^{n-1}}{n}\sum_{j=1}^{n}(-1)^{j-1}\sqrt{1-x_{j,n}^2}P_k'(x_{j,n})\frac{T_n(x)}{x-x_{j,n}}, \tag{15.3.21}$$

and get

$$\begin{aligned}\sum_{k=1}^{m}|P_k'(x)| &\le \frac{2^{n-1}}{n}\sum_{j=1}^{n}\sum_{k=1}^{m}\sqrt{1-x_{j,n}^2}|P_k'(x_{j,n})|\frac{|T_n(x)|}{|x-x_{j,n}|}\\ &\le Mn\frac{2^{n-1}}{n}\sum_{j=1}^{n}\frac{|T_n(x)|}{|x-x_{j,n}|} \le Mn^2. \end{aligned}\tag{15.3.22}$$

Here we have used inequality (15.3.19):

$$\sum_{k=1}^{m} \sqrt{1-x_{j,n}^2}|P_k'(x_{j,n})| \le Mn, \tag{15.3.23}$$

and took into account that (see [PS] problem VI. 80)

$$\frac{2^{n-1}}{n}\sum_{j=1}^{n}\frac{|T_n(x)|}{|x-x_{j,n}|} = \frac{2^{n-1}}{n}\sum_{j=1}^{n}\frac{T_n(x)}{x-x_{j,n}} = \frac{T_n'(x)}{n} \le n$$

$$\text{if} \quad \cos\left(\frac{\pi}{2n}\right) \le |x| \le 1. \tag{15.3.24}$$

Lemma 15.3.1 is proved. □

Proof of Theorem 15.3.1. Let $Q_n(x)$ be the polynomial of the best approximation for the function $f(x)$, i.e.,

$$||f - Q_n|| = E_n(f).$$

Note that

$$L_n(x, Q_n) = Q_n(x).$$

Therefore

$$\begin{aligned} ||f(x) - L_n(x,f)|| &\le ||f - Q_n|| + ||Q_n - L_n(x,Q_n)|| \\ &\quad + ||L_n(x,Q_n) - L_n(x,f)|| \\ &\le E_n(f) + \lambda_n E_n(f), \end{aligned} \tag{15.3.25}$$

where λ_n is defined in (15.3.3)

Let $P_{n-1}(x)$ be the polynomial of the best approximation for f', i.e.,

$$||f' - P_{n-1}|| = E_{n-1}(f').$$

Let

$$P_n(x) := f(0) + \int_0^x P_{n-1}(t)dt, \quad P_n' = P_{n-1}. \tag{15.3.26}$$

We have

$$L_n'(x, P_n) = P_n' = P_{n-1},$$

$$|L_n'(x,P_n) - L_n'(x,f)| \le ||f - P_n|| |L_n'(x,f), \tag{15.3.27}$$

and

$$||f - P_n|| \le ||\int_0^x |f'(t) - P_{n-1}(t)|dt|| \le E_{n-1}(f'). \qquad (15.3.28)$$

Using estimates (15.3.26) - (15.3.28), we obtain:

$$\begin{aligned} |f'(x) - L_n'(x,f)| &\le |f'(x) - P_{n-1}(x)| + |P_{n-1}(x) - L_n'(x,P_n)| \\ &\quad + |L_n'(x,P_n) - L_n'(x,f)| \\ &\le E_{n-1}(f') + \lambda_n'(x)E_{n-1}(f'), \qquad (15.3.29) \end{aligned}$$

which is inequality (15.3.3).

Lemma 15.3.1 and the definition (15.3.3) of λ_n imply

$$\lambda_n'(x) \le \frac{n\lambda_n}{\sqrt{1-x^2}}, \quad \lambda_n' \le n^2\lambda_n. \qquad (15.3.30)$$

This argument allows one to estimate

$$\lambda_n^{(k)}(x) := \sum_{j=1}^{n+1} |l_j^{(k)}(x)|$$

and

$$\lambda_n^{(k)} := ||\lambda_n^{(k)}(x)||_0.$$

For example,

$$\lambda_n^{(k)} \le n^{2k}\lambda_n.$$

Theorem 15.3.1 is proved. □

Proof of Theorem 15.3.2. We have

$$\begin{aligned} ||f(x) - L_{n,\delta}(x)|| &\le ||f(x) - L_n(x,f)|| - ||L_n(x, f - f_\delta)|| \\ &\le (1+\lambda_n)E_n(f) + \delta\lambda_n, \qquad (15.3.31) \end{aligned}$$

where we have used the estimate (15.3.25). We assume that n is large. If one minimizes the right-hand side of (15.3.31) with respect to n for a small fixed δ, then one gets the value of n, $n = n(\delta)$, for which $L_{n(\delta),\delta}(x)$ approximates $f(x)$ best.

It is known (see [A]) that if $f \in C^r(I)$ then

$$E_n(f) \le \frac{K_rM_r}{n^r}, \quad M_r = ||f^{(r)}||, \quad 1 < K_r \le \frac{\pi}{2}, \qquad (15.3.32)$$

where $r \geq 1$ is an integer. Let us denote

$$\mu_r := K_r M_r.$$

For large n minimization of the right-hand side of (15.3.31) can be done analytically. We have for large n

$$(1+\lambda_n)E_n(f) + \delta\lambda_n \asymp \frac{\mu_r \ln n}{n^r} + \delta \ln n, \tag{15.3.33}$$

where the notation $a \asymp b$ means $c_1 a \leq b \leq c_2 a$ and $c_1, c_2 > 0$ are constants independent of n.

A necessary condition for the minimizer of the right-hand side of (15.3.33) is

$$\delta = \mu_r \frac{\ln n}{n^r}\left(1 + O\left(\frac{1}{\ln n}\right)\right), \quad n \to 0. \tag{15.3.34}$$

Thus

$$n = n(\delta) = \left(\frac{r\mu_r}{\delta} \ln \frac{\mu_r}{\delta}\right)^{\frac{1}{r}}, \quad \delta \to 0. \tag{15.3.35}$$

From (15.3.31) and (15.3.35) it follows that formula (15.3.9) holds. Similarly we derive (15.3.10) and (15.3.11)

Theorem 15.3.2 is proved. □

If $f \in C^r(I)$, then

$$||f^{(k)}(x) - L_n^{(k)}(x, f)|| \leq c\omega\left(\frac{1}{n}, f^{(r)}\right) n^{k-r}(1 + n^k \ln), \quad k \leq r, \tag{15.3.36}$$

where $c > 0$ is a constant and $\omega\left(\delta, f^{(r)}\right)$ is the continuity modulus of $f^{(r)}$. Recall that if $f \in C(I)$ then

$$\omega(\delta, f) := \sup_{|x-y|\leq\delta,\ x,y\in I} |f(x) - f(y)|. \tag{15.3.37}$$

Let us prove (15.3.36). There exists a polynomial $S_n(x)$ such that

$$||f^{(k)} - S_n^{(k)}|| \leq c n^{k-r} \omega\left(\frac{1}{n}, f^{(r)}\right), \qquad k \leq r. \tag{15.3.38}$$

We have

$$S_n^{(k)} = L_n^{(k)}(x, f).$$

Therefore

$$\begin{aligned} \|f^{(k)} - L_n^{(k)}(x,f)\| &\le \|f^{(k)} - S_n^{(k)}\| + \|S_n^{(k)} - L_n^{(k)}(x,f)\| \\ &\le cn^{k-r}\omega\left(n, f^{(r)}\right) + \lambda_n^{(k)} cn^{-r}\omega\left(\frac{1}{n}, f^{(r)}\right) \\ &\le cn^{k-r}(1+n^k \ln)\omega\left(\frac{1}{n}, f^{(r)}\right). \end{aligned} \tag{15.3.39}$$

The results of this Section are of practical interest. For example, if $f = e^x, \quad x \in [-1,1]$, then

$$\|e^x - L_7(x,e^x)\| \le 10^{-7}, \quad \|e^x - L_6(x,e^x)\| \le 10^{-5}, \tag{15.3.40}$$

$$\begin{aligned} |e^x - L_7'(x,e^x)| &\le 10^{-5}, \quad |x| \le 0.98; \\ |e^x - L_7'(x,e^x)| &\le 10^{-7}, \quad 0.98 \le |x| \le 1. \end{aligned} \tag{15.3.41}$$

15.4 Other methods of stable differentiation

Let A_n be a sequence of operators which converge strongly on a set S of functions to the identity operator. For example, A_n can be Fejer, S. Bernstein, Vallee-Poussin, or other methods of approximation of f, such as a mollification:

$$\mathcal{M}_\epsilon f = \frac{1}{\epsilon}\int_I g\left(\frac{x-y}{\epsilon}\right) f(y)dy, \quad I = [-1,1]. \tag{15.4.1}$$

where

$$0 \le g(x) \in C_0^\infty(I), \quad \int_{-1}^1 g(x)dx = 1. \tag{15.4.2}$$

The function g is called the mollification kernel. Specifically, one may choose

$$g(x) = \begin{cases} ce^{-\frac{1}{1-x^2}}, & |x| < 1 \\ 0, & |x| \ge 1 \end{cases}$$

where $c = \text{const} > 0$ is chosen so that

$$\int_{-1}^1 g(x)dx = 1.$$

One has:

$$\lim_{\epsilon\to 0} \|\mathcal{M}_\epsilon f - f\| = 0, \quad \|\mathcal{M}_\epsilon f\|_{L^p(\mathbb{R})} \le \|f\|_{L^p(I)}, \quad p \ge 1, \tag{15.4.3}$$

$$\begin{aligned}&\left(\tfrac{d^m}{dx^m}\mathcal{M}_\epsilon f\right)(x) = \mathcal{M}_\epsilon\left(\tfrac{d^m}{dx^m} f\right)(x), \quad x \in (-1,1),\\ &\lim_{\epsilon\to 0}\left|\left|\tfrac{d^m}{dx^m}\mathcal{M}_\epsilon f - \tfrac{d^m}{dx^m} f\right|\right|_{L^p(\tilde{I})} = 0,\end{aligned} \tag{15.4.4}$$

where $\tilde{I} \subset I$ is an interval inside $(-1,1)$.

If $f \in C(I)$, then, taking $\epsilon = \frac{1}{h}$ and $-1 < x < 1$, one gets:

$$\begin{aligned}\mathcal{M}_\epsilon f &= n\int_{x-1}^{x+1} g(nt) f(x-t)dt\\ &= \int_{n(x-1)}^{n(x+1)} g(u) f\left(x - \frac{u}{n}\right) du \to f(x) \quad \text{as } n\to\infty.\end{aligned} \tag{15.4.5}$$

The convergence in (15.4.5) is uniform on any compact subset of I. To avoid the discussion of the convergence of $\mathcal{M}_\epsilon f$ at the end points of the interval$[-1,1]$ let us assume that $f \in C(\mathbb{R})$ is periodic with period 2.

Let us estimate f' given noisy data f_δ, $||f_\delta - f|| \le \delta$, and assuming that $f \in C^2(\mathbb{R})$. First, note that if $f \in C^2(\mathbb{R})$ and $f(x+2) = f(x)$, then (see (15.4.5)):

$$\begin{aligned}||(\mathcal{M}_\epsilon f)' - f'|| = ||\mathcal{M}_\epsilon(f') - f'|| &\le ||\int_{-1}^{1} g(u)|f\left(x-\frac{u}{n}\right) - f'(x)|du||\\ &\le \frac{M_2 c_1}{n},\end{aligned} \tag{15.4.6}$$

where

$$||f''|| \le M_2, \quad c_1 = \int_{-1}^{1} |u| g(u)^{-1} du, \quad \epsilon = \frac{1}{n}.$$

Using the inequality (15.4.6) and assuming $f \in C^2(\mathbb{R})$, $\epsilon = \frac{1}{n}$, we obtain:

$$\begin{aligned}||(\mathcal{M}_\epsilon f_\delta)' - f'|| &\le ||(\mathcal{M}_\epsilon(f_\delta) - f))'|| + ||(\mathcal{M}_\epsilon f_\delta)' - f'||\\ &\le \frac{1}{\epsilon^2}||\int_{-\infty}^{\infty} |g'\left(\frac{x-y}{\epsilon}\right)|dy|| + \frac{M_2 c_1}{n}.\end{aligned} \tag{15.4.7}$$

Note that

$$\int_{-\infty}^{\infty}\left|g'\left(\frac{x-y}{\epsilon}\right)\right| dy = \epsilon\int_{-1}^{1}|g'(t)|dt = c_2\epsilon = \frac{c_2}{n}. \tag{15.4.8}$$

Thus (15.4.7) yields

$$||(\mathcal{M}_{\frac{1}{n}} f_\delta)' - f'|| \le c_2\delta n + \frac{M_2 c_1}{n} := \eta(n,\delta). \tag{15.4.9}$$

Minimizing the right-hand side of (15.4.9) with respect to n for a fixed $\delta > 0$, one gets

$$n(\delta) = \frac{\gamma}{\sqrt{\delta}}, \qquad \gamma := \sqrt{\frac{M_2 c_1}{c_2}} \tag{15.4.10}$$

and

$$\eta(\delta) := \eta(n(\delta), \delta) = \gamma_1 \sqrt{\delta}, \quad \gamma_1 := c_2 \gamma + \frac{M_2 c_1}{\gamma}. \tag{15.4.11}$$

We have proved the following result.

Theorem 15.4.1. *Assume that*

$$f \in C^2(\mathbb{R}), \quad f(x+2) = f(x), \quad ||f_\delta - f|| \leq \delta, \quad ||f|| := ess\ sup_{x\in\mathbb{R}} |f(x)|.$$

Then the operator

$$R_\delta f_\delta := (\mathcal{M}_{\frac{1}{n(\delta)}} f_\delta)'$$

gives a stable approximation of f' provided that $n(\delta) = \frac{\gamma}{\sqrt{\delta}}$ with the error $\gamma_1 \sqrt{\delta}$, where the constants γ and γ_1 are defined in (15.4.10) and (15.4.11).

Remark 15.4.1. The method of the proof of Theorem 15.4.1 can be used for other (than mollification) methods of approximation of functions. For example, the Vallee-Poussin approximation method is

$$A_n f = \frac{(2n)!!}{(2n-1)!!} \frac{1}{2\pi} \int_{-\pi}^{\pi} f(y) \cos^{2n} \frac{x-y}{2} dy, \quad f(x+2\pi) = f(x), \tag{15.4.12}$$

yields a uniform approximation with the error

$$||A_n f - f|| \leq 3\omega \left(\frac{1}{\sqrt{n}}, f \right). \tag{15.4.13}$$

Many results on approximation of functions one finds in [A, Pow, Ti, Tik, T].

The method for stable numerical differentiation given noisy data f_δ, based on the approximating the identity sequence of operators A_n, can be summarized as follows: one constructs the formula for stable differentiation $R_\delta f_\delta = (A_{n(\delta)} f_\delta)'$ satisfies relation (15.1.2) if the sequence of operators A_n has the following properties

$$A_n f \to f, \quad (A_n f)' \to f' \quad \text{as} \quad n \to \infty, \tag{15.4.14}$$

where convergence is in L^∞ (or L^p) norm, and one has estimates of the type:

$$||(A_n f)' - f'|| \leq \omega(n), \tag{15.4.15}$$

where

$$\omega(n) \leq \frac{c}{n^b},$$

and $b > 0$ if $f \in C^\mu$ with $\mu > 1$. Moreover,

$$||(A_n f)' - (A_n f_\delta)'|| \leq ||f - f_\delta|| b(n) = \delta b(n), \tag{15.4.16}$$

where $\lim_{n\to\infty} b(n) = \infty$. The $n(\delta)$ is found by minimizing the quantity:

$$\delta b(n) + \omega(n) = \min, \tag{15.4.17}$$

or by solving the equation

$$\delta = \frac{\omega(n)}{b(n)} \tag{15.4.18}$$

for n. As $\delta \to 0$ one has $n(\delta) \to \infty$.

Consider the operator

$$R_h f = \frac{f(x+h) - f(x-h)}{2h} \tag{15.4.19}$$

on the space of $C^2(\mathbb{R})$ periodic functions, $f(x+2) = f(x)$, as an integral operator

$$R_h f = \int_{-1}^{1} \frac{\delta(y-h) - \delta(y+h)}{2h} f(x+y)dy \tag{15.4.20}$$

with the distributional kernel. One has

$$\lim_{h\to 0} \frac{\delta(y-h) - \delta(y+h)}{2h} = -\delta'(y) \tag{15.4.21}$$

in the sense of distributions. Thus

$$\lim_{h\to 0} R_h f = -\int_{-1}^{1} \delta'(y) f(x+y)dy = f'(x). \tag{15.4.22}$$

One may use other kernels to approximate $f'(x)$. For example, consider the operator

$$T_h f := \int_{-1}^{1} w(yh) f(x+yh)dy, \tag{15.4.23}$$

where w is an entire function. Assuming that $f \in C^2(-1,1)$ we have

$$T_h f = \int_{-1}^{1} \left[w(0) + yhw'(0) + O(y^2h^2)\right] \left[f(x) + yhf'(x) + O(y^2h^2)\right] dy. \tag{15.4.24}$$

Let us require

$$w(0) = 0, \quad w'(0) := c_1, \quad w(y) = c_1 y,$$

then (15.4.24) yields

$$T_h f = c_1 h^2 \int_{-1}^{1} y^2 dy f'(x) + O\left(h^3\right) = \frac{2h^2}{3} c_1 f'(x) + O(h^3). \tag{15.4.25}$$

Choose $c_1 = \frac{3}{2h^2}$, i.e., $w(yh) = \frac{3y}{2h}$. Then

$$\lim_{h\to 0} T_h f = f'(x). \tag{15.4.26}$$

Therefore formula (15.4.23) with $w(yh) = \frac{3y}{2h}$ gives an approximation to $f'(x)$. This formula requires an integration, so it is more complicated than formula (15.4.19). In Section 15.1 we have proved that operator (15.4.19) with $h = h(\delta)$, defined in (15.1.8), is optimal in the sense (15.1.5), i.e. among all linear and nonlinear operators acting from $L^\infty_\pi(\mathbb{R})$ into $L^\infty_\pi(\mathbb{R})$, where $L^\infty_\pi(\mathbb{R})$ is the space of 2-periodic functions on $\mathbb{R}$ with the norm $||f||_{L^\infty_\pi(\mathbb{R})} = \text{ess sup}_{x\in\mathbb{R}}|f(x)|$, the operator $R_h(\delta)$, defined in (15.4.19), with $h = h(\delta)$, defined in (15.1.8), gives the best possible approximation of f' in the sense (15.1.5).

15.5 DSM and stable differentiation

Consider the problem of stable differentiation as the problem of solving Volterra integral equation of the first kind:

$$Au := \int_0^x u dt = f(x), \quad 0 \le x \le 1. \tag{15.5.1}$$

Without loss of generality we assume that f is defined on the interval $[0,1]$ and $f(0) = 0$. If $f(0) \neq 0$, then the function $f(x) - f(0)$ vanishes at $x = 0$ and has the same derivative as f.

Let us assume that noisy data f_δ are given, $||f_\delta - f|| \le \delta$, and the norm is $L^2(0,1) := H$ norm. The Hilbert space H is assumed *real-valued.*

We have

$$(Au, u) = \int_0^1 dx \int_0^x u(t)dtu(x) = \frac{1}{2}\left(\int_0^x u(t)dt\right)^2\Bigg|_0^1 = \frac{1}{2}\left(\int_0^1 u(t)dt\right)^2 \geq 0. \quad (15.5.2)$$

Since A is a linear bounded operator in H, conditions (1.3.2) hold. We have

$$||Au||^2 = \int_0^1 dx\left(\int_0^x udt\right)^2 dx \leq \int_0^1 x\int_0^x u^2dtdx \leq \frac{1}{2}||u||^2.$$

Thus $||A|| \leq \frac{1}{\sqrt{2}}$. Consider a DSM scheme for solving equation (15.5.1):

$$\dot{u}_\delta(t) = -A_{a(t)}^{-1}[Au_\delta(t) + a(t)u_\delta(t) - f_\delta] = -u_\delta(t) + A_{a(t)}^{-1} f_\delta,$$
$$u_\delta(0) = u_0, \quad (15.5.3)$$

where $a(t)$ satisfies (6.1.30). Inequality (15.5.2) and the linearity of A imply

$$(Au - Av, u - v) \geq 0, \quad \forall u, v \in H. \quad (15.5.4)$$

Thus, we may use Theorems 6.2.1 and 6.3.1 for calculating a stable approximation to f' given the noisy data f_δ.

Calculating A_a^{-1} amounts to solving the problem

$$\int_0^x u(y)dy + au(x) = g, \quad g \in H. \quad (15.5.5)$$

This problem is solved analytically:

$$u := u_a(x) = \frac{d}{dx}\frac{1}{a}\int_0^x e^{-\frac{x-y}{a}} g(y)dy - \frac{g}{a} - \frac{1}{a^2}\int_0^x e^{-\frac{x-y}{a}} g(y)dy. \quad (15.5.6)$$

If $a = a(t)$, formula (15.5.6) remains valid. One can use formula (15.5.6) for constructing a stable approximation of f' given noisy data f_δ. This corresponds to solving the equation

$$Au_{a,\delta} + au_{a,\delta} = f_\delta, \quad (15.5.7)$$

and choosing $a = a(\delta)$ so that

$$u_\delta := u_{a(\delta),\delta}$$

would converge to f':

$$\lim_{\delta\to 0} ||u_\delta - f'|| = 0. \tag{15.5.8}$$

To choose $a(\delta)$ we take $g = f_\delta$ in (15.5.6) and estimate the difference $u_{a,\delta} - f'$:

$$||u_{a,\delta} - f'|| \le ||u_{a,\delta} - u_a|| + ||u_a - f'||, \tag{15.5.9}$$

where u_a is the right-hand side of (15.5.6) with f in place of g. Using (15.5.6), we get

$$||u_{a,\delta} - u_a|| \le \frac{\delta}{a} + \delta ||\frac{1}{a^2}\int_0^x e^{-\frac{x-y}{a}}(f_\delta - f)dy|| \le \delta\left(\frac{1}{a} + \frac{1}{a^{\frac{3}{2}}\sqrt{2}}\right), \tag{15.5.10}$$

where $||f_\delta - f|| \le \delta$. Furthermore, integrating by parts, we get

$$\begin{aligned}
||u_a - f'|| &= ||\frac{f}{a} - \frac{1}{a^2}\int_0^x e^{-\frac{x-y}{a}} f(y)dy - f'|| \\
&\le ||\frac{f}{a} - \frac{1}{a^2}\left[ae^{-\frac{x-y}{a}} f\Big|_0^x - a\int_0^x e^{-\frac{x-y}{a}} f'(y)dy\right] - f'|| \\
&\le ||\frac{1}{a}f(0) + \frac{1}{a}\int_0^x e^{-\frac{x-y}{a}} f'(y)dy - f'(x)|| \\
&\le ||\frac{1}{a}\int_0^x e^{-\frac{s}{a}} f'(x-s)ds - f'(x)||.
\end{aligned} \tag{15.5.11}$$

Here we have used the assumption

$$f(0) = 0.$$

We have

$$\lim_{a\to 0} ||\frac{1}{a}\int_0^x e^{-\frac{s}{a}} f'(x-s)ds - f'(x)|| = 0. \tag{15.5.12}$$

If $T(a) : H \to H$ is defined as

$$T(a)h = \frac{1}{a}\int_0^x e^{-\frac{x-y}{a}} h(y)dy, \tag{15.5.13}$$

then

$$\begin{aligned}
||T(a)h||^2 &= \int_0^1 dx \left|\frac{1}{a}\int_0^x e^{-\frac{x-y}{a}} h(y)dy\right|^2 \\
&\le \int_0^1 dx \frac{1}{a^2}\int_0^x e^{-\frac{2(x-y)}{a}} dy \int_0^x h^2 dy \\
&\le \frac{1}{a^2}\int_0^1 dx \frac{a}{2}\int_0^1 h^2 dy = \frac{1}{2a}||h||^2.
\end{aligned}$$

Thus

$$||T(a)|| \leq \frac{1}{\sqrt{2a}}. \tag{15.5.14}$$

We have

$$\begin{aligned} ||T(a)h - h|| &= ||\frac{1}{a}\int_0^x e^{-\frac{x-y}{a}}[h(y) - h(x)]dy - e^{-\frac{x}{a}}h(x)|| \\ &\leq ||\frac{1}{a}\int_0^x e^{-\frac{s}{a}}[h(x-s) - h(x)]ds|| \\ &\quad + ||e^{-\frac{x}{a}}h(x)||. \end{aligned} \tag{15.5.15}$$

Assume that

$$\sup_{0\leq x\leq 1} |h^{(j)}(x)| \leq M_j, \quad 0 \leq j \leq 2, \quad M_j = \text{const}. \tag{15.5.16}$$

Then (15.5.15) implies:

$$||T(a)h - h|| \leq M_1 a + M_0 \left(\frac{a}{2}\right)^{\frac{1}{2}}, \tag{15.5.17}$$

and (15.5.11) implies

$$||u_a - f'|| \leq M_2 a + M_1 \left(\frac{a}{2}\right)^{\frac{1}{2}}. \tag{15.5.18}$$

If $h(x)$ in (15.5.15) is an arbitrary element of $H = L^2(0,1)$ then by the Lebesgue's dominated convergence theorem we have

$$\lim_{a\to 0} ||e^{-\frac{x}{a}}h(x)|| = 0, \tag{15.5.19}$$

and, setting $\frac{s}{a} = z$, we obtain:

$$\begin{aligned} &\lim_{a\to 0} ||\frac{1}{a}\int_0^x e^{-\frac{s}{a}}[h(x-s) - h(x)]ds||^2 \\ =\ &\lim_{a\to 0} ||\int_0^{\frac{x}{a}} e^{-z}[h(x-az) - h(x)]dz||^2 \\ \leq\ &\lim_{a\to 0} \int_0^1 dx \int_0^{\frac{1}{a}} e^{-z}dz \int_0^{\frac{1}{a}} e^{-z}|h(x-az) - h(x)|^2 dz \\ =\ &0. \end{aligned} \tag{15.5.20}$$

Therefore

$$\lim_{a\to 0} ||T(a)h - h|| = 0, \quad \forall h \in H. \tag{15.5.21}$$

This conclusion is a consequence of (15.5.17) as well.

It follows from (15.5.9), (15.5.10) and (15.5.18) that

$$||u_{a,\delta} - f'|| \le \frac{\delta}{a^{\frac{3}{2}}\sqrt{2}}\left(1 + \sqrt{2}a^{\frac{1}{2}}\right) + \frac{M_1 a^{\frac{1}{2}}}{\sqrt{2}}\left(1 + \frac{M_2 a^{\frac{1}{2}}}{M_1}\sqrt{2}\right). \quad (15.5.22)$$

Minimizing the right-hand side of (15.5.22) with respect to a we get

$$a_\delta = O\left(\delta^{\frac{1}{2}}\right). \quad (15.5.23)$$

Denoting $u_\delta := u_{a_\delta,\delta}$ and using (15.5.22), we obtain:

$$||u_\delta - f'|| = O\left(\delta^{\frac{1}{2}}\right), \quad \delta \to 0. \quad (15.5.24)$$

Numerical results for stable differentiation, based on the equation (15.5.7), are given in [ARU].

Let us formulate the result we have proved.

Theorem 15.5.1. *Assume (15.5.16), and let*

$$u_\delta = T(a(\delta))f_\delta,$$

where $||f_\delta - f|| \le \delta$, T_a *is defined in (15.5.13) and* $a(\delta) = O\left(\delta^{\frac{1}{2}}\right)$. *Then (15.5.24) holds.*

Let us return to DSM (15.5.3) and give the stopping rule, i.e. the choice of t_δ such that

$$\lim_{\delta \to 0} ||u_\delta(t_\delta) - f'|| = 0. \quad (15.5.25)$$

The solution to (15.5.3) is

$$u_\delta(t) = u_0 e^{-t} + \int_0^t e^{-(t-s)} A_{a(s)}^{-1} f_\delta ds := u(t, f_\delta). \quad (15.5.26)$$

Thus

$$||u_\delta(t) - f'|| \le ||u(t, f_\delta) - u(t, f)|| + ||u(t, f) - f'|| \le \frac{\delta}{a(t)} + ||u(t, f) - f'||. \quad (15.5.27)$$

We have

$$||u(t, f) - f'|| \le ||u_0||e^{-t} + ||\int_0^t e^{-(t-s)} A_{a(s)}^{-1} f ds - f'||. \quad (15.5.28)$$

From (15.5.18) it follows that

$$||A_{a(s)}^{-1} f - f'|| = O\left(a^{\frac{1}{2}}\right), \quad a(s) \to 0. \tag{15.5.29}$$

Since

$$\lim_{t\to\infty} \int_0^t e^{-(t-s)} h(s) ds = h(\infty),$$

provided $h(\infty)$ exists, we conclude from (15.5.28) and (15.5.29) that

$$||u(t,f) - f'|| = O\left(a^{\frac{1}{2}}(t)\right), \quad t \to \infty. \tag{15.5.30}$$

From (15.5.27) and (15.5.30) we obtain

$$||u_\delta(t) - f'(x)|| \le \frac{\delta}{a(t)} + O\left(a^{\frac{1}{2}}(t)\right). \tag{15.5.31}$$

Thus, if t_δ minimizes (15.5.31) then $\lim_{\delta\to 0} t_\delta = \infty$ and (15.5.25) holds. The minimizer a_δ of the right-hand side of (15.5.31) with respect to a is $a_\delta = O\left(\delta^{\frac{2}{3}}\right)$. Therefore t_δ is found from the relation $a_\delta = a(t)$, and

$$||u_\delta(t_\delta) - f'(x)|| \le O\left(\delta^{\frac{1}{3}}\right), \quad \text{as} \quad \delta \to 0. \tag{15.5.32}$$

Consider now another DSM:

$$\dot{u}_\delta = -(Au_\delta + a(t)u_\delta - f_\delta), \quad u(0) = 0. \tag{15.5.33}$$

Our arguments are valid for an arbitrary initial data $u(0) = u_0$, and we took $u_0 = 0$ just for simplicity of writing.

Denote by

$$u_\delta(t) = u(t, f_\delta)$$

the solution to (15.5.33), and by $V(t)$ the solution to the equation

$$AV + a(t)V - f_\delta = 0. \tag{15.5.34}$$

Let

$$u_\delta - V(t) := w. \tag{15.5.35}$$

Then

$$\dot{w} = -\dot{V} - [Aw + a(t)w], \quad w(0) = 0. \tag{15.5.36}$$

We have proved above (see (15.5.24)) that:

$$\lim_{\delta \to 0} ||V(t_\delta) - f'|| = 0, \tag{15.5.37}$$

where t_δ is defined by the equation $a_\delta = a(t)$ and $\lim_{\delta \to 0} t_\delta = \infty$. Thus, (15.5.25) holds if we prove that

$$\lim_{\delta \to 0} w(t_\delta) = 0. \tag{15.5.38}$$

Multiply (15.5.36) by w, denote

$$g(t) := ||w(t)||,$$

and use the inequality $(Aw, w) \geq 0$ to get:

$$\dot{g} \leq -a(t)g + ||\dot{V}||. \tag{15.5.39}$$

Let us estimate $||\dot{V}||$. Differentiate (15.5.34) with respect to t and get

$$A\dot{V} + a(t)\dot{V} = -\dot{a}V. \tag{15.5.40}$$

Multiply (15.5.40) by $\dot{V}$, use inequality $(A\dot{V}, \dot{V}) \geq 0$, and get

$$||\dot{V}|| \leq \frac{|\dot{a}|}{a(t)} ||V||. \tag{15.5.41}$$

Similarly, multiply (15.5.34) by V and get

$$||V|| \leq \frac{||f_\delta||}{a(t)}.$$

This and (15.5.4) imply

$$||\dot{V}|| \leq \frac{|\dot{a}|}{a^2} ||f_\delta||.$$

Assume

$$\lim_{t \to \infty} \frac{|\dot{a}(t)|}{a^3(t)} = 0, \quad \int_0^\infty a(t)dt = \infty. \tag{15.5.42}$$

From (15.5.39) we obtain:

$$g(t) = g_0 e^{-\int_0^t a(s)ds} + e^{-\int_0^t a(s)ds} \int_0^t e^{\int_0^s a(p)dp} ||\dot{V}(s)|| ds := J_1 + J_2.$$

From the second condition (15.5.42) it follows that

$$\lim_{t\to\infty} ||g_0||e^{-\int_0^t a(s)ds} = 0.$$

We have

$$J_2(t) \leq \frac{\int_0^t e^{\int_0^s a(p)dp} \frac{|\dot{a}(s)|}{a^2(s)} ds ||f_\delta||}{e^{\int_0^t a(s)ds}}. \quad (15.5.43)$$

Applying L'Hospital's rule to (15.5.43) and using the first condition (15.5.42), we conclude that

$$\lim_{t\to\infty} J_2(t) = 0.$$

Thus,

$$\lim_{t\to\infty} g(t) = 0. \quad (15.5.44)$$

Let us summarize the result.

Theorem 15.5.2. *Assume that $f \in C^2$, $0 < a(t) \searrow 0$, (15.5.42) holds, and $||f_\delta - f||_{L^2(0,1)} \leq \delta$. Then problem (15.5.33) has a unique global solution $u_\delta(t)$, and (15.5.25) holds with t_δ chosen from the equation $a_\delta = a(t)$, where a_δ is given in (15.5.23).*

Remark 15.5.1. One can construct an iterative method for stable numerical differentiation using the general results from Sections 2.4 and 4.4.

15.6 Stable calculating singular integrals

Let us denote the hypersingular integral by the symbol

$$⨍_0^b f(x)dx.$$

We assume that

$$f \in C^k([0, b]), \quad k > \mu - 1 > 0, \quad b > 0,$$

and define

$$⨍_0^b \frac{f(x)dx}{x^\mu} := \int_0^b \frac{f(x) - f_k(x)}{x^\mu} dx + \sum_{j=0}^{k} \frac{f^{(j)}(0) b^{-\mu+j+1}}{j!(-\mu + j + 1)}, \quad (15.6.1)$$

where

$$f_k(x) := \sum_{j=0}^{k} \frac{f^{(j)}(0)x^j}{j!}. \tag{15.6.2}$$

If $k \geq \mu - 1$ and $f \in C^{k+1}([0,b])$, then the term in (15.6.1), corresponding to $\mu = j+1$ is replaced by $\frac{f^{(j)}(0)\ln b}{j!}$. If the integral is given on the interval $[a,d]$, it can be reduced to the integral over $[0,b]$ by a change of variables. The definition (15.6.1) does not depend on the choice of k in (15.6.1) in the region $k > \mu - 1$. Although there are many papers on calculating hypersingular integrals, paper [R18] was the first one, as far as the author knows, in which the ill-possedness of this problem was discussed and a stable approximation of hypersingular integrals was proposed. We present here the results of this paper. Let us first explain why the usual quadrature formulas lead to possibly very large errors in computing hypersingular integrals.

In other words, we explain why computing hypersingular integrals is an ill-posed problem.

Consider a quadrature formula

$$Q_n f = \sum_{j=1}^{n} w_{n,j} f(x_{n,j}), \tag{15.6.3}$$

where $w_{n,j}$ are the weights and $x_{n,j}$ are the nodes. Let us assume that

$$\lim_{n\to\infty} Q_n f = \int_0^b f dx \quad \forall f \in C([0,b]). \tag{15.6.4}$$

If one applies formula (15.6.3) to calculating the right-hand side of (15.6.1), and if the function f is given with some error, so that f_δ is given in place of f,

$$||f_\delta(x) - f(x)|| \leq \delta, \quad ||\cdot|| = ||\cdot||_{L^\infty(0,b)}, \tag{15.6.5}$$

then, even if we assume that $f^{(j)}(0), \quad 0 \leq j \leq k$ are known exactly, we have

$$\lim_{n\to\infty} \sup_{\{f_\delta : ||f_\delta - f|| \leq \delta\}} \left| Q_n \left[\frac{f_\delta(x) - f_k(x)}{x^\mu}\right] - Q_n \left[\frac{f(x) - f_k(x)}{x^\mu}\right]\right| = \infty. \tag{15.6.6}$$

Thus, the problem of calculating hypersingular integrals is ill-posed. Let us check (15.6.6). Take, for example,

$$f_\delta(x) = \begin{cases} f(x) + \delta, & \text{if } 0 < a \leq x \leq b, \\ f(x) + \delta\left(\frac{x}{a}\right)^\mu, & 0 < x \leq a. \end{cases} \tag{15.6.7}$$

Then the function $f_\delta(x) - f(x)$ is continuous on $[0, b]$, and

$$\lim_{n\to\infty} \sum_{j=1}^{n} w_{n,j} \frac{[f_\delta(x_{n,j}) - f(x_{n,j})]}{x_{n_j}^{\mu}} = \delta \int_0^a \left(\frac{x}{a}\right)^{\mu} \frac{dx}{x^{\mu}} + \delta \int_a^b dx > \delta \frac{1}{a^{\mu-1}}. \quad (15.6.8)$$

The right-hand side of (15.6.8) tends to infinity as $a \to 0$, so that (15.6.6) is verified.

The problem is to construct a quadrature formula for stable computation of integral (15.6.1) given noisy data f_δ. We will not assume tht $f^{(j)}(0)$ are known, but estimate them stably from noisy data f_δ. This can be done by the method developed in Section 5.1 provided that upper bounds on the derivatives of f are known. To estimate stably $f^{(j)}(0), \quad 0 \le j \le k$, we use the bounds

$$||f^{(j)}|| := \sup_{0\le x\le b} |f^{(j)}(x)| := M_j, \quad 0 \le j \le k+2. \quad (15.6.9)$$

One may look for an estimate of $f^{(j)}(0)$ of the form:

$$f^{(j)}(0) = \sum_{m=0}^{j} c_{m,j} f(mh) := L_{j,h} f, \quad (15.6.10)$$

where $c_{m,j}$ are found from the condition

$$f^{(j)}(0) - \sum_{m=0}^{j} c_{m,j} f(mh) = O\left(h^{j+1}\right). \quad (15.6.11)$$

This leads to a linear algebraic system for finding $c_{m,j}$:

$$\sum_{m=0}^{j} \frac{m^p h^p}{p!} = \delta_{jp}, \quad 0 \le p \le j. \quad (15.6.12)$$

This system is uniquely solvable because its matrix has non-zero Vandermonde determinant.

If f_δ is given in place of f,

$$||f_\delta - f|| \le \delta,$$

then one uses formula (15.6.10) with f_δ in place of f and finds $h = h(\delta)$ such that

$$\lim_{\delta\to 0} [L_{j,h(\delta)} f_\delta - f^{(j)}(0)] = 0. \quad (15.6.13)$$

This $h(\delta)$ is found from the estimate:

$$[L_{j,h}f_\delta - f^{(j)}(0)| \leq |L_{j,h}(f_\delta - f)| + |L_{j,h}f - f^{(j)}(0)| \leq \delta\varphi_j(h) + O\left(h^{j+1}\right), \tag{15.6.14}$$

where

$$\varphi_j(h) = \sum_{m=0}^{j} |c_{m,j}|.$$

The coefficients

$$c_{m,j} = c_{m,j}(h)$$

and

$$\varphi_j(h) \to \infty \quad \text{as} \quad h \to 0.$$

One finds $h(\delta)$ by minimizing the right-hand side of (15.6.14) with respect to h for a given $\delta > 0$. This gives also an estimate of the error of the approximation (15.6.10).

Assume now that stable approximations $f_\delta^{(j)}(0)$ to $f^{(j)}(0)$, $0 \leq j \leq k$, have been calculated. Denote

$$f_{k\delta}(x) := \sum_{j=0}^{k} \frac{f_\delta^{(j)}(0)}{j!} x^j, \tag{15.6.15}$$

and let

$$Q_\delta f_\delta := Q_{n,a,b}\left(\frac{f_\delta(x) - f_{k\delta}(x)}{x^\mu}\right) + \sum_{j=0}^{k} \frac{f_\delta^{(j)}(0) b^{-\mu+j+1}}{j!(-\mu+j+1)}, \tag{15.6.16}$$

Here $Q_{n,a,b}$, $n = 1, 2, \cdots$, are quadrature formulas with positive weights which converge, as $n \to \infty$, to $\int_a^b f(x)dx$ for any $f \in C([a,b])$.

For example, quadrature formulas of Gaussian type have these properties, they are exact for polynomials of degree $\leq 2n - 1$ and converge for any $f \in C([a,b])$ (see, e.g., [DR]).

Let us formulate our result.

Theorem 15.6.1. *Assume $k > \mu - 1$, $f \in C^{k+2}([0,b])$. Then there is an $a = a(\delta, k)$ such that*

$$J := \left| \int_0^b f dx - Q_\delta f_\delta \right| = O\left(\delta^{\frac{k+2-\mu}{k+1}}\right), \qquad \delta \to 0, \tag{15.6.17}$$

for all $n \geq n(\delta, k, \mu, f)$.

Proof. Denote

$$J_1 = \int_0^a \frac{f(x) - f_k(x)}{x^\mu} dx, \quad J_1 = O\left(a^{k+2-\mu}\right), \tag{15.6.18}$$

$$J_2 = \int_a^b \frac{f(x) - f_k(x)}{x^\mu} dx - Q_{n,a,b}\left(\frac{f(x) - f_k(x)}{x^\mu}\right) \to 0 \quad \text{as} \quad n \to \infty, \tag{15.6.19}$$

$$\begin{aligned} J_3 &= Q_{n,a,b}\left(\frac{f(x) - f_\delta(x)}{x^\mu}\right), \\ |J_3| &\leq O\left(\delta a^{-\mu+1}\right) \quad \text{as} \quad n \to \infty, \end{aligned} \tag{15.6.20}$$

$$J_4 = Q_{n,a,b}\left(\frac{f_{k\delta}(x) - f_k(x)}{x^\mu}\right) + \sum_{j=0}^{k} \frac{f^{(j)}(0) - f_\delta^{(j)}(0)}{j!} \frac{b^{-\mu+j+1}}{(-\mu+j+1)}, \tag{15.6.21}$$

$$J_4 = O\left(\delta^{\frac{k+2-\mu}{k+1}}\right), \quad \text{as} \quad n \to \infty. \tag{15.6.22}$$

From these estimates the relation (15.6.17) follows.

Theorem 15.6.1 is proved. □

Chapter 16

Auxiliary results from analysis

16.1 Contraction mapping principle

Let F be a mapping in a Banach space X. Assume that there is a closed set $D \subset X$ such that

$$F(D) \subset D, \quad ||F(u)-F(v)|| \leq q||u-v||, \quad u, v \in D, \quad q \in (0,1). \tag{16.1.1}$$

Theorem 16.1.1. *If (16.1.1) holds then equation*

$$u = F(u) \tag{16.1.2}$$

has a unique solution in D. This solution can be obtained by the iterative method

$$u_{n+1} = F(u_n), \quad u_0 \in D, \quad u = \lim_{n\to\infty} u_n, \tag{16.1.3}$$

where $u_0 \in D$ is arbitrary, and

$$||u_n - u|| \leq \frac{q^n}{1-q}||u_1 - u_0||. \tag{16.1.4}$$

Proof. One has

$$||u_{n+1}-u_n|| = ||F(u_n)-F(u_{n-1})|| \leq q||u_n-u_{n-1}|| \leq \cdots \leq q^n||u_1-u_0||. \tag{16.1.5}$$

Thus

$$\begin{aligned} ||u_n - u_p|| &\leq \sum_{j=p+1}^{n} ||u_j - u_{j-1}|| \\ &\leq \sum_{j=p+1}^{n} q^{j-1}||u_1 - u_0|| \leq \frac{q^p}{1-q}||u_1 - u_0||. \end{aligned} \tag{16.1.6}$$

By the Cauchy test there exists the limit

$$\lim_{n\to\infty} u_n = u, \tag{16.1.7}$$

and

$$||u - u_n|| \le \frac{q^n}{1-q}||u_1 - u_0||.$$

Passing to the limit in (16.1.3) yields (16.1.2). Uniqueness of the solution to (16.1.2) in D is immediate: if u and v solve (16.1.2) then

$$||u - v|| = ||F(u) - F(v)|| \le q||u - v||. \tag{16.1.8}$$

If $0 < q < 1$, then (16.1.8) implies $||u - v|| = 0$, so $u = v$.
Theorem 16.1.1 is proved. □

Remark 16.1.1. If

$$||F^m(u) - F^m(v)|| \le q||u - v||, \quad u, v \in D, \quad q \in (0,1), \tag{16.1.9}$$

where $m > 1$ is an integer, and $F(D) \subset D$, then equation (16.1.2) has a unique solution in D.

Indeed, the equation

$$u = F^m(u) \tag{16.1.10}$$

has a unique solution in D by Theorem 16.1.1 due to assumption (16.1.9). Therefore $F(u) = F^{m+1}(u) = F^m(F(u))$. Since the solution to (16.1.10) is unique, it follows that $u = F(u)$, so u solves equation (16.1.2). The solution to (16.1.2) is unique if assumption (16.1.9) holds. Indeed, if $v = F(v)$, then $v = F^m(v)$, and therefore $v = u$, because the solution to (16.1.10) is unique. One can also prove that (16.1.3) holds.

Remark 16.1.2. The contraction assumption (16.1.1) cannot be replaced by a weaker one

$$||F(u) - F(v)|| < ||u - v||. \tag{16.1.11}$$

For example, if $X = \mathbb{R}^1$, $D = X$ and $F(u) = \frac{\pi}{2} + u - \arctan u$ then $u = F(u)$ implies $\frac{\pi}{2} = \arctan u$, and there is no real number u such that $\arctan u = \frac{\pi}{2}$. On the other hand,

$$|F(u) - F(v)| = |u - v - (\arctan u - \arctan v)| < |u - v|.$$

Remark 16.1.3. If F maps D into a precompact subset of D and (16.1.11) holds, then equation (16.1.2) has a solution in D and this solution is unique.

Uniqueness of the solution is obvious: if v and u solve (16.1.2), then, by (16.1.11) we have:

$$||u - v|| = ||F(u) - F(v)|| < ||u - v||. \tag{16.1.12}$$

Thus, $u = v$.

Let us prove the existence of the solution to (16.1.2). Consider F on $F(D)$. The set $F(D)$ is precompact by the assumption. It is closed because F is continuous and D is closed. Thus the set $F(D)$ is is compact.

The continuous function $||u - F(u)||$ on the compact set $F(D)$ attains its minimum at some point y, i.e.

$$||y - F(y)|| \le ||u - F(u)|| \quad \forall u \in F(D). \tag{16.1.13}$$

If $||y - F(y)|| > 0$, then

$$||F(y) - F^2(y)|| < ||y - F(y)||,$$

which is a contradiction to (16.1.13). Therefore $y = F(y)$, and equation (16.1.2) has a solution.

Remark 16.1.4. Assume that F depends continuously on a parameter $z \in Z$, where Z is a Banach space, i.e.

$$\lim_{||z-\xi|| \to 0} ||F(u, z) - F(u, \xi)|| = 0 \quad \forall u \in D \subset X. \tag{16.1.14}$$

Consider the equation

$$u = F(u, z). \tag{16.1.15}$$

Theorem 16.1.2. *Assume that for every*

$$z \in B(z_0, \epsilon) := \{z : z \in Z, \quad ||z - z_0|| \le \epsilon\}, \quad \epsilon = const > 0,$$

one has

$$||F(u, z) - F(v, z)|| \le q||u - v||, \tag{16.1.16}$$

where $q \in (0, 1)$ *does not depend on* $z \in \overline{B(z_0, \epsilon)}$,

$$\lim_{||z-z_0|| \to 0} ||F(u, z) - F(u, z_0)|| = 0 \quad \forall u \in D, \tag{16.1.17}$$

and $F(\cdot, z)$ maps a closed set $D \subset X$ into itself for every $z \in B(z_0, \epsilon)$. Then equation (16.1.15) has a unique solution $u(z)$, which is continuous as $z \to z_0$:

$$\lim_{||z-z_0||\to 0} ||u(z) - u(z_0)|| = 0. \tag{16.1.18}$$

If $F(\cdot, z) \in C^m(B(z_0, \epsilon))$, then $u \in C^m(B(z_0, \epsilon))$, $\quad m \geq 1$.

Proof. For each $z \in B(z_0, \epsilon)$ equation (16.1.15) has a unique solution $u(z)$. Moreover

$$\begin{aligned} ||u(z) - u(z_0)|| &= ||F(u(z), z) - F(u(z_0), z_0)|| \\ &\leq ||F(u(z), z) - F(u(z_0), z)|| \\ &\quad + ||F(u(z_0), z) - F(u(z_0), z_0)|| \\ &\leq q||u(z) - u(z_0)|| + ||F(u(z_0), z) - F(u(z_0), z_0)||. \end{aligned}$$

Therefore

$$||u(z) - u(z_0)|| \leq \frac{1}{1-q}||F(u(z_0), z) - F(u(z_0), z_0)||. \tag{16.1.19}$$

By assumption (16.1.17) one gets from (16.1.19) the desired conclusion (16.1.18). If $m = 1$, then differentiate (16.1.15) with respect to z and get the equation for $u'_z := u'$:

$$u' = F_u(u, z)u' + F_z(u, z).$$

From (16.1.16) it follows that

$$||F'_u(u, z)|| \leq q < 1,$$

so that the above equation has a unique solution and this implies $u \in C^1(B(z_0, \epsilon))$. Similarly one treats the case $m > 1$.

Theorem 16.1.2 is proved. □

Remark 16.1.5. Under the assumption of Theorem 16.1.1 the sequence u_n, defined in (16.1.3), is a minimizing sequence for the functional $g(u) := ||u - F(u)||$.

Remark 16.1.6. Suppose that there is a norm $||\cdot||$ on X equivalent to the original norm, i.e.

$$c_1||u||_1 \leq ||u|| \leq c_2||u||_1 \quad \forall u \in X. \tag{16.1.20}$$

It may happen that the map F is not a contraction on D with respect to the original norm, but is a contraction with respect to an equivalent norm. A standard example deals with the equation

$$u(t) = \int_0^t f(s, u(s))ds := F(u), \tag{16.1.21}$$

where $u(t)$ is a continuous vector-function with values in $\mathbb{R}^n$. Assume that

$$|f(t,u) - f(t,v)| \leq k|u - v|, \tag{16.1.22}$$

where

$$|u| := \left(\sum_{j=1}^{n} |u_j|^2\right)^{\frac{1}{2}}$$

is the length of the vector u, and $k > 0$ is a constant independent of u, v and t. Let X be the space $C(0,T)$ of continuous vector-functions with the norm

$$||u|| = \max_{0 \leq t \leq T} |u(t)|. \tag{16.1.23}$$

Define an equivalent norm:

$$||u||_1 = \max_{0 \leq t \leq T} \{e^{-\gamma t}|u(t)|\}, \quad \gamma = \text{const} > 0. \tag{16.1.24}$$

We have

$$\begin{aligned} ||F(u) - F(v)|| &\leq \max_{0 \leq t \leq T} \left\{ e^{-\gamma t} k \int_0^t |u - v| ds \right\} \\ &\leq \max_{0 \leq t \leq T} \left\{ e^{-\gamma t} k \int_0^t e^{\gamma s} ds \right\} ||u - v||_1 \\ &\leq k \frac{1 - e^{-\gamma T}}{\gamma} ||u - v||_1. \end{aligned} \tag{16.1.25}$$

One can always choose $\gamma > 0$ such that

$$q := k \frac{1 - e^{-\gamma T}}{\gamma} < 1 \tag{16.1.26}$$

no matter how large the fixed k and T are. It is easy to see that in the original norm (corresponding to $\gamma = 0$) the map F is not necessarily a contraction if k and T are sufficiently large.

16.2 Existence and uniqueness of the local solution to the Cauchy problem

Let

$$\dot{u} = F(t,u), \quad u(0) = u_0, \tag{16.2.1}$$

where $F : X \to X$ is a map in a Banach space. Assume that

$$||F(t,u) - F(t,v)|| \leq M_1(R)||u - v||, \quad u, v \in B(u_0, R), \tag{16.2.2}$$

where

$$B(u_0, R) = \{u : ||u - u_0|| \leq R, \quad u \in X\},$$

and

$$||F(t,u)|| \leq M_0(R), \quad \forall u, v \in B(u_0, R). \tag{16.2.3}$$

For any fixed $u \in B(u_0, R)$ the element $F(t,u)$ is assumed continuous with respect to $t \in [0,T]$. Consider the equation

$$u(t) = u_0 + \int_0^t F(s, u(s))ds := G(u) \tag{16.2.4}$$

in the space $C([0,T]); X)$ of continuous functions with values in X and the norm

$$|u| = \max_{0 \leq t \leq T} ||u(t)||. \tag{16.2.5}$$

Let us check that the map G is a contraction on the set

$$D = \{u(t) : u \in B(u_0, R), \quad t \in [0,T]\},$$

provided that T is sufficiently small, and $GD \subset D$.

Using (16.2.3), we get

$$|u(t) - u_0| \leq M_0(R)T \leq R \quad \text{if} \quad T \leq \frac{R}{M_0(R)}, \tag{16.2.6}$$

so

$$GD \subset D \quad \text{if} \quad T \leq \frac{R}{M_0(R)}. \tag{16.2.7}$$

Furthermore, using (16.2.2), we get

$$|G(u) - G(v)| \leq M_1(R)T|u - v| := q|u - v|, \tag{16.2.8}$$

where

$$q = M_1(R)T < 1 \qquad \text{if} \qquad T \leq \frac{R}{M_1(R)}. \tag{16.2.9}$$

From Theorem 16.1.1 we obtain the following result.

Theorem 16.2.1. *Assume (16.2.2), (16.2.3), (16.2.6) and (16.2.9). Then problem (16.2.1) has a unique solution $u = u(t, u_0) \in B(u_0, R)$ defined on $[0, T]$, where*

$$T = \min\left(\frac{1}{M_1(R)}, \frac{R}{M_0(R)}\right). \tag{16.2.10}$$

Remark 16.2.1. By Theorem 16.1.2 the solution $u(t, u_0)$ depends continuously on the initial approximation u_0 in the following sense.

If

$$||u_0 - v_0|| \leq \epsilon, \quad T = \min\left(\frac{1}{M_1(R+\epsilon)}, \frac{R}{M_0(R+\epsilon)}\right),$$

then

$$\lim_{||v_0 - u_0|| \to 0} |u(t, u_0) - u(t, v_0|) = 0.$$

Theorem 16.2.1 allows one to introduce the notion of maximal interval of the existence of the solution. Namely, if the solution $u(t)$ exists on the interval $[0, T)$ and does not exist on the interval $[0, \tau]$ for any $\tau > T$, then we say that $[0, T]$ is the maximal interval of the existence of the solution to (16.2.1). We have defined the maximal interval of the form $[0, T)$ of the existence of the solution. Usually the definition does not fix the lower point of the maximal interval of the existence of the solution, which is zero in our case (see, e.g., [H] for the standard definition of the maximal interval of the existence of the solution).

Under the assumptions of Theorem 16.2.1 we can prove that if $[0, T]$ is the maximal interval of the existence of the solution to (16.2.1) then

$$\lim_{t \to T^-} ||u(t)|| = \infty, \tag{16.2.11}$$

where $t \to T^-$ denotes convergence from the left.

Indeed, assuming that (16.2.11) fails, we have

$$\sup_{0\le t\le T} ||u(t)|| \le c < \infty. \tag{16.2.12}$$

Thus

$$||u(t_n)|| \le c, \quad t_n \to T, \quad t_n < T. \tag{16.2.13}$$

Consider the problem

$$u(t) = u(t_n) + \int_{t_n}^{t} F(s, u(s))ds, \tag{16.2.14}$$

and let

$$D = \{u(t) : u(t) \in B(u(t_n), R), \quad t \in [t_n, t_n + \tau]\}.$$

We choose t_n such that

$$t_n > T - \tau \tag{16.2.15}$$

and

$$\tau < \min\left(\frac{1}{M_1(R)}, \frac{1}{M_0(R)}\right), \tag{16.2.16}$$

where

$$||u(t) - u(t_n)|| \le R, \quad t_n \le t \le t_n + \tau. \tag{16.2.17}$$

If the constant c, independent of n, is given in (16.2.13), then we can find R and $\tau = \tau(R) > 0$, independent of n, so that the inequalities (16.2.15) and (16.2.17) hold. Therefore the unique solution $u(t)$ to problem (16.2.1) is defined on the interval $[0, T_1]$, where $T_1 = t_n + \tau > T$. This contradicts to the assumption that $T < \infty$ and $[0, T]$ is the maximal interval of the existence of the solution $u(t)$. We have proved the following result.

Theorem 16.2.2. *Assume that conditions (16.2.2) and (16.2.3) hold with some $R > 0$ for any u_0 such that $||u_0|| \le c$, where $c > 0$ is an arbitrary constant and $R = R(c)$. Then either the maximal interval $[0, T]$ of the existence of the solution to (16.2.1) is finite, and then (16.2.11) holds, or it is infinite, and then (16.2.12) holds for any $T < \infty$ with a constant $c = c(T) > 0$.*

Remark 16.2.2. Theorems 16.2.1 and 16.2.2 imply that a unique local solution to problem (16.2.1) is a global one provided that a uniform with respect to time a priori estimate

$$\sup_t ||u(t)|| \leq c < \infty \tag{16.2.18}$$

is established for the solution to problem (16.2.1) and $F(t, u)$ satisfies the assumptions of Theorem 16.2.2.

This argument was used often in Chapters 3, 6-11, 13.

Remark 16.2.3. We give two standard examples of the finite maximal interval of the existence of the solution to problem (16.2.1).

Example 16.2.1. Let

$$\dot{u} = 1 + u^2, \quad u(0) = 1. \tag{16.2.19}$$

Then

$$u = \tan\left(\frac{\pi}{4} + t\right). \tag{16.2.20}$$

Thus, $u(t)$ does not exist on any interval of length greater than π, and

$$\lim_{t \to \frac{\pi}{4}} u(t) = \infty. \tag{16.2.21}$$

Example 16.2.2. Let

$$u_t - \Delta u = u^2, \quad t \geq 0, \quad x \in D \subset R^n, \tag{16.2.22}$$

$$u_N|_S = 0, \quad t \geq 0, \tag{16.2.23}$$

$$u|_{t=0} = u_0(x). \tag{16.2.24}$$

Here, D is a bounded domain with a sufficient smooth boundary S, N is the exterior normal to S, and

$$u_0(x) \geq 0, \quad \int_D u_0(x)dx > 0. \tag{16.2.25}$$

Let us verify that the local solution to problem (16.2.22) - (16.2.25) has to blow up in a finite time, that is (16.2.11) holds for some $T < \infty$. Indeed, integrate (16.2.22) over D, use formula

$$\int_D \Delta u dx = \int_S u_N ds = 0,$$

and get

$$\frac{\partial}{\partial t} \int_D u dx = \int_D u^2 dx. \tag{16.2.26}$$

Moreover, assuming that u is real-valued, we get

$$g(t) := \int_D u dx \le \left(\int_D u^2 dx \right)^{\frac{1}{2}} |D|^{\frac{1}{2}}, \quad |D| := \text{meas } D, \tag{16.2.27}$$

where meas D is the volume of the domain D.

Thus,

$$\frac{dg}{dt} \ge cg^2, \quad c := \frac{1}{|D|}, \quad g(0) = \int_D u_0 dx > 0. \tag{16.2.28}$$

Integrate (16.2.28) and get

$$g(t) \ge \frac{1}{\frac{1}{g(0)} - ct}. \tag{16.2.29}$$

Therefore

$$\lim_{t \to T^-} g(t) = +\infty, \quad T := \frac{1}{cg(0)}. \tag{16.2.30}$$

If (16.2.30) holds, then

$$\lim_{t \to T} ||u||_{L^2(D)} = \infty, \tag{16.2.31}$$

because of the estimate (16.2.27).

16.3 Derivatives of nonlinear mappings

Let $F : X \to Y$ be a mapping from a Banach space X into a Banach space Y.

Definition 16.3.1. *If there exists a bounded linear map $A = A(u)$, such that*

$$F(u+h) = F(u) + A(u)h + o(||h||), \quad ||h|| \to 0, \tag{16.3.1}$$

then the map F is called F-differentiable (Fréchet differentiable) at the point u and one writes $A(u) := F'(u)$. If $A(u)$ depends continuously on u in the sense

$$\lim_{||v-u||} ||A(u) - A(v)|| = 0, \quad u, v \in D, \tag{16.3.2}$$

then we write $F \in C^1(D)$. Here D is an open set in X, and $||A(u)||$ is the norm of the linear operator from X to Y.

If $F'(u)$ exists, then it is unique.

Definition 16.3.2. *The map $F : X \to Y$ is called G-differentiable (Gâteaux differentiable) at a point u if*

$$F(u + th) = F(u) + tA(u)h + o(t), \quad t \to 0, \tag{16.3.3}$$

where t is a number, $h \in X$ is any arbitrary element, and $A(u)$ is a bounded linear operator $A(u) : X \to Y$.

The term $o(t)$ in (16.3.3) depends on h. Clearly if F is F-differentiable then F is G-differentiable.

If F is F-differentiable at a point u, then F is continuous at this point. One has

$$F(u + h) - F(u) = \int_0^1 \frac{d}{dt} F(u + th)dt = \int_0^1 F'(u + th)hdt.$$

Thus

$$||F(u + h) - F(u)|| \le \sup_{v \in B(u,R)} ||F'(v)||||h||, \quad ||h|| \le R,$$

where $B(u, R) : \{u : ||u - v|| \le R\}$. The first derivative $F'(u)$ is a map $X \to L(X, Y)$, where $L(X, Y)$ is the space of bounded linear operators from X into Y. If this map is F-differentiable, then its derivative is called the second derivative of F, $F''(u)$. The map $F''(u)$ maps X into a set $L(X, L(X, Y))$. This set can be identified with the set of $L(X, X; Y)$ of bilinear mappings. An m-linear mapping $G : X_1 \times X_2 \times \cdots \times X_m \to Y$ is a mapping which is a bounded linear mapping with respect to each of the variables x_j assuming that all the other variables are fixed, thus

$$||G(x_1, x_2, \cdots, x_m)|| \le c \prod_{j=1}^{m} ||x_j||.$$

One can define higher order derivatives inductively. Then $F^{(n)}(u)$ is the first derivative of $F^{(n-1)}(u)$. If $F \in C^{n+1}(D)$, i.e., F is $n + 1$ times F-differentiable in an open set $D \subset X$, then an analog of Taylor's formula holds:

$$F(u + h) = \sum_{j=0}^{n} \frac{F^{(j)}(u)h^j}{j!} + R_n, \tag{16.3.4}$$

where

$$R_n = \int_0^1 \frac{(1 - s)^n}{n!} F^{(n+1)}(u + sh)h^{n+1}ds. \tag{16.3.5}$$

Here

$$F^{(n)}(u)h^n = \frac{\partial^n}{\partial t_1 \cdots \partial t_n} F\left(u + \sum_{j=1}^{n} t_j h_j\right)\Bigg|_{\substack{t_1 = \cdots = t_n = 0 \\ h_1 = \cdots = h_n = h}} . \quad (16.3.6)$$

Example 16.3.1. Uryson operators are defined by the formula:

$$F(u) = \int_D K(x, y, u(y))dy, \quad D \subset \mathbb{R}^n. \quad (16.3.7)$$

They are considered in $X = C(D)$ or $X = L^p(D)$, where $C(D)$ is the space of continuous in a bounded domain D the function with sup-norm and $L^p(D)$, $p \geq 1$, are the Lebesgue spaces. If the function $K(x, y, u(y))$ is continuous in the region $D \times D \times B_R$, where $B = \{u : |u| < R\}$, and $K_u = \frac{\partial K}{\partial u}$ is continuous in $D \times D \times B_R$, then the operator (16.3.7) is F-differentiable in $C(D)$ and in $L^p(D)$ at any point u such that $||u|| \leq R$. One has

$$F'(u)h = \int_D K_u(x, y, u(y))h(y)dy. \quad (16.3.8)$$

If one wishes that the operator F, defined in (16.3.7), be defined on all of the space $X = C(D)$, then the function $K(x, y, u)$ has to be defined in $D \times D \times \mathbb{R}$. If $X = L^p(D)$, then it is necessary to restrict the growth of K with respect to variable u in order that the domain of definition of F in X be nontrivial.

For example, for F, defined in (16.3.7), to act from $L^{p_1}(D)$ int $L^{p_2}(D)$, it is sufficient that

$$|K(x, y, u)| \leq c_1 + c_2|u|^{\frac{p_1}{p_2}},$$

where c_1 and c_2 are positive constants. For F, defined in (16.3.7) to act in $L^p(D)$, it is sufficient that

$$|K(x, y, u)| \leq a(x, y)(c_1 + c_2|u|^\gamma),$$

$$\int_D \int_D |a(x, y)|^b dxdy < \infty, \quad \gamma \leq b - 1,$$

$$\frac{\gamma b}{b-1} \leq p \leq b,$$

where c_1, c_2, γ, and b are positive constants. This is verified by using Hölder inequality.

A particular case of the Uryson operator (16.3.7) is the Hammerstein operator

$$F(u) = \int_D K(x,y)f(y,u(y))dy, \tag{16.3.9}$$

where the function $f(y,u)$ is continuous with respect to u and integrable with respect to y. The operator

$$(Nu)(y) := f(y,u(y))$$

is called Nemytskij operator. It acts from $L^{p_1}(D)$ into $L^{p_2}(D)$ if

$$|f(y,u(y))| \leq a(y) + b|u(y)|^{\frac{p_1}{p_2}}, \quad a \in L^{p_2}(D). \tag{16.3.10}$$

The operator

$$Ku := \int_D K(x,y)u(y)dy \tag{16.3.11}$$

acts from $L^p(D)$ into $L^q(D)$, $\quad p, q \geq 1$, if

$$\int_D \int_D |K(x,y)| dx dy < \infty, \quad r' = \frac{r}{r-1}, \quad r := \min(p, q'). \tag{16.3.12}$$

Indeed

$$|Ku| \leq \left(\int_D |K(x,y)|^{r'}\right)^{\frac{1}{q'}} \left(\int_D |u(y)|^q\right)^{\frac{1}{q'}}, \tag{16.3.13}$$

and $r \leq q'$ implies $r' \geq q$. Note that since $|D| :=$ meas $D < \infty$ and $r \leq p$, one has

$$||u||_{L^r(D)} \leq ||u||_{L^p(D)}. \tag{16.3.14}$$

Furthermore

$$\begin{aligned}\left(\int_D |Ku|^q dx\right)^{\frac{1}{q}} &< \left(\int_D dx \left(\int_D |K(x,y)|^{r'} dy\right)^{\frac{q}{r'}}\right)^{\frac{1}{q}} ||u||_{L^p(D)} \\ &\leq \left(\int_D \int_D |K(x,y)|^{r'} dy dx\right)^{\frac{1}{r'}} \times \\ &\qquad |D|^{\frac{1}{q\left(\frac{r'}{q}\right)}} ||u||_{L^p(D)}.\end{aligned} \tag{16.3.15}$$

From (16.3.13) - (16.3.15) the desired conclusion follows and

$$||K||_{L^p(D)\to L^q(D)} \leq ||K(x,y)||_{L^{r'}(D\times D)} |D|^{\frac{1}{q\left(\frac{r'}{q}\right)'}}. \tag{16.3.16}$$

16.4 Implicit function theorem

Assume that X, Y and Z are Banach spaces $U \subset X$ and $V \subset Y$ are open sets, $u_0 \in U$, $v_0 \in V$, $F : X \times Y \to Z$ is a $C^1(U \times V)$ map, and $F(u_0, v_0) = 0$.

Theorem 16.4.1. *If* $[F'_v(u_0, v_0)]^{-1} := L$ *is a bounded linear operator, then there exists a unique map* $v = f(u)$, *such that* $F(u, f(u)) = 0$, $u \in U_1 \subset U$, $v_0 = f(u_0)$, f *is continuous in* U_1. *If* $F \in C^m(U \times V)$ *then* $f \in C^m(U_1)$.

Proof. Without loss of generality assume that $u_0 = 0$, $v_0 = 0$. Consider the equation

$$v = v - LF(u, v) := T(v, u). \tag{16.4.1}$$

Let us check that the operator T maps a ball $B_\epsilon := \{v : ||v|| \leq \epsilon\}$ into itself and is a contraction on B_R for any $u \in U_1$. If this is checked, then Theorem 16.1.1 implies the existence and uniqueness of the solution to $v = f(u)$ to the equation (16.4.1), and Theorem 16.1.2 implies that $f \in C^m(U_1)$ if $F \in C^m(U \times V)$.

We have

$$\begin{aligned} ||v - LF(u, v)|| &= ||v - LF_u(0, 0)u - v - LR|| \\ &\leq ||L|| ||F_u(0, 0)|| ||u|| + ||L|| o(||v|| + ||u||), \end{aligned} \tag{16.4.2}$$

where $R = R(u, v)$ is the remainder in the formula

$$F(u, v) = F_u(0, 0)u + F_v(0, 0)v + R, \quad R = o(||u|| + ||v||), \quad F_v(0, 0) = L^{-1}. \tag{16.4.3}$$

If

$$||u|| \leq \delta, \quad ||L|| ||F_u(0, 0)|| \delta \leq \frac{\epsilon}{2}, \tag{16.4.4}$$

and

$$o(||v|| + ||u||) \leq \frac{\epsilon}{2}, \tag{16.4.5}$$

then

$$T(v, u) B_\epsilon \subset B_\epsilon \quad \forall u \in \{u : ||u|| \leq \delta\} := \tilde{B}_\delta \subset U. \tag{16.4.6}$$

Let $v, w \in B_\epsilon$ and $u \in \tilde{B}_\delta$. Then

$$T'_v(v, u) = I - LF'_v(u, v), \tag{16.4.7}$$

and, by the continuity of $F_v'(u,v)$,

$$||F_v'(u,v) - F_v(0,0)|| \leq \alpha(\delta,\epsilon), \tag{16.4.8}$$

where

$$\lim_{\delta,\epsilon\to 0} \alpha(\delta,\epsilon) = 0. \tag{16.4.9}$$

Since $LF_v'(0,0) = I$, formulas (16.4.7) - (16.4.9) imply

$$||T_v'(v,u)|| \leq ||L|| ||F_v'(u,v) - F_v'(0,0)|| \leq q < 1, \tag{16.4.10}$$

provided that

$$0 \leq \epsilon \leq \epsilon_0, \quad 0 \leq \delta \leq \delta_0, \tag{16.4.11}$$

where ϵ_0 and δ_0 are sufficiently small. The number $q < 1$ in (16.4.10) depends on ϵ_0 and δ_0 but not on $u \in \tilde{B}_{\epsilon_0}$ or $v \in \tilde{B}_{\delta_0}$.

Thus, Theorems 16.1.1 and 16.1.2 imply the conclusions of Theorem 16.4.1. □

A particular case of Theorem 16.4.1 is the equation for the inverse function

$$v = f(u), \quad v_0 = f(u_0). \tag{16.4.12}$$

Theorem 16.4.2. *Assume that $f : Y \to X$,*

$$L := [f'(v_0)]^{-1}, \tag{16.4.13}$$

is a bounded linear operator, $u_0 = f(v_0)$, and $f \in C^1(V)$, $\quad v_0 \in V$, where V is an open set in Y. Then there exists and is unique the inverse function $u = f^{-1}(v)$, such that $u = f^{-1}(f(u))$, $\quad u \in U$, $\quad u_0 \in U$, where $U \subset X$ is a neighborhood of u_0, and $f^{-1}(v)$ is C^m if $f \in C^m$.

Proof. Let $F(u,v) = f(v) - u$, then

$$F_v(u_0, v_0) = f'(v_0), \tag{16.4.14}$$

the operator $f'(v_0)$ has a bounded inverse operator (16.4.13), and Theorem 16.4.2 follows from Theorem 16.4.1. □

16.5 An existence theorem

Consider the Cauchy problem:

$$\dot{u} = F(u,t), \quad u(0) = u_0. \tag{16.5.1}$$

Let

$$J = [0,a] \subset \mathbb{R}, \quad \Lambda = [\alpha, \beta] \subset \mathbb{R}_+,$$

where X_λ is a scale of Banach spaces, satisfying the following assumptions:

$$X_\lambda \subset X_\mu \quad \text{if } \mu < \lambda, \quad \lambda, \mu \in \Lambda, \tag{16.5.2}$$

$$||u||_\mu \leq ||u||_\lambda \quad \text{if } \mu < \lambda, \quad \lambda, \mu \in \Lambda. \tag{16.5.3}$$

Theorem 16.5.1. *Assume that*

$$F : J \times X_\lambda \to X_\mu \quad \textit{is continuous if} \quad \mu < \lambda, \quad \lambda, \mu \in \Lambda, \tag{16.5.4}$$

$$F(t,0) \in X_\beta, \tag{16.5.5}$$

$$||F(t,u) - F(t,v)||_\mu \leq \frac{M}{\lambda - \mu} ||u - v||_\lambda, \quad \mu < \lambda, \quad t \in J \quad \lambda, \mu \in \Lambda. \tag{16.5.6}$$

Then problem (16.5.1) with $u_0 \in X_\beta$ has a unique solution $u(t) \in X_\lambda$ for every $\lambda \in (\alpha, \beta)$, and this solution is defined on the interval $t \in [0, \delta(\beta - \lambda))$, where $\delta := \min\left(a, \frac{1}{Me}\right)$.

Proof. Let

$$u_{n+1}(t) = u_0 + \int_0^t F(s, u_n(s))ds. \tag{16.5.7}$$

Assumptions (16.5.2) - (16.5.6) imply that $u_n(t) \in X_\lambda$ for every $\lambda \in [\alpha, \beta]$, and the map $u_n : J \to X_\lambda$ is continuous.

Denote $J(t) = [0,t]$ and

$$m(t) := ||u_0||_\beta + \frac{\beta - \alpha}{M} \max_{s \in J(t)} ||F(s,0)||_\beta. \tag{16.5.8}$$

Let us check that

$$||u_{n+1}(t) - u_n(t)||_\lambda \leq m(t) \left(\frac{tMe}{\beta - \lambda}\right)^{n+1}, \quad t \in J. \tag{16.5.9}$$

We have

$$\begin{aligned} ||u_1(t)|| - u_0(t)||_\lambda &\le t\left[\frac{M}{\beta-\lambda}||u_0||_\beta + F(s,0)||_\beta\right] \\ &\le \frac{Mt}{\beta-\lambda}m(t). \end{aligned} \tag{16.5.10}$$

If (16.5.9) holds for $n = j-1$, then it holds for $n = j$. Indeed, with $\gamma < \beta - \lambda$ we have:

$$\begin{aligned} ||u_{j+1}(t) - u_j(t)||_\lambda &\le \int_0^t ||F(s,u_j(s)) - F(s,u_{j-1}(s))||_\lambda ds \\ &\le \frac{M}{\gamma}\int_0^t ||u_j(s) - u_{j-1}(s)||_{\lambda+\gamma} ds \\ &\le \frac{M}{\gamma}\int_0^t m(s)\left(\frac{sMe}{\beta-\lambda-\gamma}\right)^j ds. \end{aligned} \tag{16.5.11}$$

Choose

$$\gamma = \frac{\beta-\lambda}{j+1}. \tag{16.5.12}$$

Then

$$\begin{aligned} ||u_{j+1}(t)|| - u_j(t)||_\lambda &\le \frac{M(j+1)}{\beta-\lambda}m(t)\frac{(Me)^j t^{j+1}}{(\beta-\lambda)^j\left(1-\frac{1}{j+1}\right)^j (j+1)} \\ &= m(t)\left(\frac{tMe}{\beta-\lambda}\right)^{j+1}\frac{1}{e\left(1-\frac{1}{j+1}\right)^j} \\ &\le m(t)\left(\frac{tMe}{\beta-\lambda}\right)^{j+1}, \end{aligned} \tag{16.5.13}$$

because

$$\left(1+\frac{1}{j}\right)^j \le e. \tag{16.5.14}$$

Thus (16.5.9) is proved.

If $t \in [0, \delta(\beta-\lambda))$, where $\delta = \min\left(a, \frac{1}{Me}\right)$, then

$$\frac{tMe}{\beta-\lambda} := q < 1. \tag{16.5.15}$$

Therefore (16.5.9) implies

$$\lim_{n\to\infty} u_n(t) = u(t), \tag{16.5.16}$$

where convergence is in X_λ uniform in $t \in [0, \delta(\beta-\lambda))$, and $u(t)$ solves the equation

$$u(t) = u_0 + \int_0^t F(s, u(s))ds \tag{16.5.17}$$

and problem (16.5.1).

If $v(t)$ is another solution to (16.5.1) in X_λ, then $w := u(t) - v(t)$ solves the problem:

$$\dot{w} = F(t,u) - F(t,v), \quad w(0) = 0. \tag{16.5.18}$$

Let

$$\begin{aligned} u_{n+1}(t) &= u_0 + \int_0^t F(s, u_n(s))ds, \\ v_{n+1}(t) &= v_0 + \int_0^t F(s, v_n(s))ds, \\ w_n(t) &= u_n(t) - v_n(t). \end{aligned}$$

Then, as in (16.5.13), we have:

$$||w_n(t)||_\lambda \le m(t)\left(\frac{tMe}{\beta-\lambda}\right),$$

and, if (16.5.15) holds, then

$$\lim_{n\to\infty} ||w_n(t)||_\lambda = 0. \tag{16.5.19}$$

Thus $w(t) = 0$ for $t \in [0, t_0]$, where $t_0 < \delta(\beta - \lambda)$.

Theorem 16.5.1 is proved. □

16.6 Continuity of solutions to operator equations with respect to a parameter

Let X and Y be Banach spaces, $k \in \Delta \subset \mathbb{C}$ be a parameter, Δ be an open bounded set on a complex plane $\mathbb{C}$, $A(k) : X \to Y$ be a map, possibly nonlinear, $f := f(k) \in Y$ be a function.

Consider an equation

$$A(k)u(k) = f(k). \tag{16.6.1}$$

We are interested in conditions, sufficient for the continuity of $u(k)$ with respect to $k \in \Delta$. The novel points in our presentation include necessary and sufficient conditions for the continuity of the solution to equation (16.6.1) with linear operator $A(k)$ and sufficient conditions for its continuity when the operator $A(k)$ is nonlinear.

Consider separately the cases when $A(k)$ is a linear map and when $A(k)$ is a nonlinear map.

Assumptions 1. $A(k) : X \to Y$ is a linear bounded operator, and

(a) equation (16.6.1) is uniquely solvable for any

$$k \in \Delta_0 := \{k : |k - k_0| \le r\}, \quad k_0 \in \Delta, \quad \Delta_0 \subset \Delta,$$

(b) $f(k)$ is continuous with respect to $k \in \Delta_0$, $\sup_{k\in\Delta_0} \|f(k)\| \le c_0$;

(c) $\lim_{h\to 0} \sup_{\substack{k\in\Delta_0 \\ v\in M}} \|[A(k+h) - A(k)]v\| = 0$, where $M \subset X$ is an arbitrary bounded set,

(d) $\sup_{\substack{k\in\Delta_0 \\ f\in N}} \|A^{-1}(k)f\| \le c_1$, where $N \subset Y$ is an arbitrary bounded set and c_1 may depend on N.

Theorem 16.6.1. *If **Assumptions 1** hold, then*

$$\lim_{h\to 0} \left\|u(k+h) - u(k)\right\| = 0. \tag{16.6.2}$$

Proof. One has

$$\begin{aligned} u(k+h) - u(k) &= A^{-1}(k+h)f(k+h) - A^{-1}(k)f(k) \\ &= A^{-1}(k+h)f(k+h) - A^{-1}(k)f(k+h) \\ &\quad + A^{-1}(k)f(k+h) - A^{-1}(k)f(k). \end{aligned} \tag{16.6.3}$$

Furthermore,we have:

$$\left\|A^{-1}(k)\big[f(k+h) - f(k)\big]\right\| \le c_1\left\|f(k+h) - f(k)\right\| \to 0 \text{ as } h \to 0. \tag{16.6.4}$$

Using the relation:

$$A^{-1}(k+h) - A^{-1}(k) = -A^{-1}(k+h)\big[A(h+k) - A(k)\big]A^{-1}(k),$$

and estimating the norms of inverse operators, we obtain

$$\|A^{-1}(k+h) - A^{-1}(k)\| \le c_1^2 \|A(k+h) - A(k)\| \to 0 \text{ as } h \to 0. \quad (16.6.5)$$

From (16.6.3) - (16.6.5) and **Assumptions 1** the conclusion of Theorem 16.6.1 follows. □

Remark 16.6.1. **Assumptions 1** are not only sufficient for the continuity of the solution to (16.6.1), but also necessary if one requires the continuity of $u(k)$ uniform with respect to f running through arbitrary bounded sets. Indeed, the necessity of the assumption (a) is clear; that of the assumption (b) follows from the case $A(k) = I$, where I is the identity operator; that of the assumption (c) follows from the case $A(k) = I$, $A(k+h) = 2I$, $\forall h \neq 0$, $f(k) = g \neq 0$ $\forall k \in \Delta_0$. Indeed, in this case assumption c) fails and one has $u(k) = g$, $u(k+h) = \frac{g}{2}$, so $||u(k+h) - u(k)|| = \frac{||g||}{2}$ does not tend to zero as $h \to 0$.

To prove the necessity of the assumption (d), suppose that

$$\sup_{k \in \Delta_0} \|A^{-1}(k)\| = \infty.$$

Then, by the Banach-Steinhaus theorem, there is an element f such that $\sup_{k \in \Delta_0} \|A^{-1}(k) f\| = \infty$, so that

$$\lim_{j \to \infty} \|A^{-1}(k_j) f\| = \infty, \qquad k_j \to k \in \Delta_0.$$

Then

$$\|u_j\| := \|u(k_j)\| = \|A^{-1}(k_j) f\| \to \infty,$$

so u_j does not converge to $u := u(k) = A^{-1}(k) f$, although $k_j \to k$.

Assumptions 2. $A(k) : X \to Y$ is a nonlinear map, and (a), (b), (c) and (d) of **Assumptions 1** hold, and the following assumption holds:

(e) $A^{-1}(k)$ *is a homeomorphism of* X *onto* Y *for each* $k \in \Delta_0$.

Remark 16.6.2. Assumption (e) is included in (d) in the case of a linear operator $A(k)$ because if $\|A(k)\| \le c_2$ and $\|A^{-1}(k)\| \le c_1$ then $A(k)$, $k \in \Delta_0$, is an isomorphism of X onto Y.

Theorem 16.6.2. *If **Assumptions** 2 hold, then (16.6.2) holds.*

Let us make the following **Assumption** A_d:
Assumptions A_d: **Assumptions 2** hold and

(f) $\dot{f}(k) := \frac{df(k)}{dk}$ is continuous in Δ_0,

(g) $\dot{A}(u,k) := \frac{\partial A(u,k)}{\partial k}$ is continuous with respect to (wrt) k in Δ_0 and wrt $u \in X$,

(j) $\sup_{k\in\Delta_0} \|[A'(u,k)]^{-1}\| \leq c_3$, where $A'(u,k)$ is the Fréchet derivative of $A(u,k)$ and $[A'(u,k)]^{-1}$ is continuous with respect to u and k. $\dot{f}(k) := \frac{df(k)}{dk}$ is continuous in Δ_0.

Remark 16.6.3. If **Assumption** A_d holds, then

$$\lim_{h\to 0} \|\dot{u}(k+h) - \dot{u}(k)\| = 0. \tag{16.6.6}$$

Remark 16.6.4. If **Assumptions 1** hold except one: $A(k)$ is not necessarily a bounded linear operator, $A(k)$ may be unbounded, closed, densely defined operator-function, then the conclusion of Theorem 16.6.2 still holds and its proof is the same. For example, let

$$A(k) = L + B(k),$$

where $B(k)$ is a bounded linear operator continuous with respect to $k \in \Delta_0$, and L is a closed, linear, densely defined operator from $D(L) \subset X$ into Y. Then

$$\|A(k+h) - A(k)\| = \|B(k+h) - B(k)\| \to 0 \quad \text{as} \quad h \to 0,$$

although $A(k)$ and $A(k+h)$ are unbounded.

Proofs of Theorem 16.6.2 and of Remark 16.6.3 are given below.

Proof of Theorem 16.6.2. One has:

$$A(k+h)u(k+h) - A(k)u(k) = f(k+h) - f(k) = o(1) \quad \text{as} \quad h \to 0.$$

Thus

$$A(k)u(k+h) - A(k)u(k) = o(1) - [A(k+h)u(k+h) - A(k)u(k+h)].$$

Since

$$\sup_{\{u(k+h):\|u(k+h)\|\leq c\}} \|A(k+h)u(k+h) - A(k)u(k+h)\| \underset{h\to 0}{\to} 0,$$

one gets

$$A(k)u(k+h) \to A(k)u(k) \text{ as } h \to 0. \tag{16.6.7}$$

By the **Assumptions 2**, item (e), the operator $A(k)$ is a homeomorphism. Thus (16.6.7) implies (16.6.2).

Theorem 16.6.2 is proved. □

Proof of Remark 16.6.3. First, assume that $A(k)$ is linear. Then

$$\frac{d}{dk}A^{-1}(k) = -A^{-1}(k)\dot{A}(k)A^{-1}(k), \quad \dot{A} := \frac{dA}{dk}. \tag{16.6.8}$$

Indeed, differentiate the identity $A^{-1}(k)A(k) = I$ and get

$$\frac{dA^{-1}(k)}{dk}A(k) + A^{-1}(k)\dot{A}(k) = 0.$$

This implies (16.6.8).

This argument proves also the existence of the deriviative $\frac{dA^{-1}(k)}{dk}$. Formula $u(k) = A^{-1}(k)f(k)$ and the continuity of $\dot{f}$ and of $\frac{dA^{-1}(k)}{dk}$ yield the existence and continuity of $\dot{u}(k)$. Remark 16.6.3 is proved for linear operators $A(k)$.

Assume now that $A(k)$ is nonlinear, $A(k)u := A(k,u)$. Then one can differentiate (16.6.1) with respect to k and get

$$\dot{A}(k,u) + A'(k,u)\dot{u} = \dot{f}, \tag{16.6.9}$$

where A' is the Fréchet derivative of $A(k,u)$ with respect to u. Formally one assumes that $\dot{u}$ exists, when one writes (16.6.9), but in fact (16.6.9) proves the existence of $\dot{u}$, because $\dot{f}$ and $\dot{A}(k,u) := \frac{\partial A(k,u)}{\partial k}$ exist by the **Assumption** A_d and $[A'(k,u)]^{-1}$ exists and is an isomorphism by the **Assumption** A_d, item (j). Thus, (16.6.9) implies

$$\dot{u} = [A'(k,u)]^{-1}\dot{f} - [A'(k,u)]^{-1}\dot{A}(k,u). \tag{16.6.10}$$

Formula (16.6.10) and **Assumption** A_d imply (16.6.6).

Remark 16.6.3 is proved. □

Consider some application of the above results to Fredholm equations depending on a parameter.

Let

$$Au := u - \int_D b(x,y,k)u(y)dy := [I - B(k)]u = f(k), \tag{16.6.11}$$

where $D \subset R^n$ is a bounded domain, $b(x,y,k)$ is a function on $D \times D \times \Delta_0$, $\Delta_0 := \{|k - k_0| < r\}$, $k_0 > 0$, $r > 0$ is a sufficiently small number. Assume that $A(k_0)$ is an isomorphism of $H := L^2(D)$ onto H, for example,

$\int_D \int_D |b(x,y,k_0)|^2 dxdy < \infty$ and $\mathcal{N}(I - B(k_0)) = \{0\}$, where $\mathcal{N}(A)$ is the null-space of A. Then, $A(k_0)$ is an isomprohism of H onto H by the Fredholm alternative, and **Assumptions 1** hold if $f(k)$ is continuous with respect to $k \in \Delta_0$ and

$$\lim_{h\to 0} \int_D \int_D |b(x,y,k+h) - b(x,y,k)|^2 dx\, dy = 0 \quad k \in \Delta_0. \qquad (16.6.12)$$

Condition (16.6.12) implies that if $A(k_0)$ is an isomorphism of H onto H, then so is $A(k)$ for all $k \in \Delta_0$ if $|k - k_0|$ is sufficiently small.

Remark 16.6.3 applies to (16.6.11) if $\dot{f}$ is continuous with respect to $k \in \Delta_0$, and $\dot{b} := \frac{\partial b}{\partial k}$ is continuous with respect to $k \in \Delta_0$ as an element of $L^2(D \times D)$. Indeed, under these assumptions one has

$$\dot{u} = [I - B(k)]^{-1}(\dot{f} - \dot{B}(k)u),$$

and the right-hand side of this formula is continuous in Δ_0.

16.7 Monotone operators in Banach spaces

Suppose that $F : X \to X^*$ is a map from a real Banach space X into its adjoint.

Definition 16.7.1. *X is uniformly convex if $||u|| = ||v|| = 1$ and $||u-v|| \geq \epsilon$ imply $||\frac{u+v}{2}|| \leq 1 - \delta, \quad \forall \epsilon (0,2]$ and $\delta = \delta(\epsilon) > 0$.*

X is locally uniformly convex if $||u|| = ||v|| = 1$ and $||u - v|| \geq \epsilon$ imply $||\frac{u+v}{2}|| \leq 1 - \delta, \quad \forall \epsilon (0,2]$ and $\delta = \delta(u, \epsilon) > 0$.

X is strictly convex if $||u|| = ||v|| = 1$ and $u \neq v$ implies

$$||su + (1-s)v|| < 1 \quad \forall s \in (0,1).$$

Theorem 16.7.1. *a) If X is uniformly convex , then X is reflexive and locally uniformly convex.*

b) If X is locally uniformly convex, then X is strictly convex and

$$\{u_n \rightharpoonup u, \quad ||u_n|| \to ||u||\} \quad \textit{implies} \quad u_n \to u.$$

c) X is strictly convex if and only if $||u+v|| = ||u|| + ||v||$ implies $v = 0$ or $u = \lambda v, \quad \lambda = const \geq 0$.

d) X is strictly convex if and only if every $u \in X, \quad u \neq 0$, considered as an element of X^{**}, attains its norm at exactly one element $u^* \in X^*, \quad ||u|| = \sup_{||u^*||=1} |u^*(u)|$.

e) X is reflexive if and only if every $x^* \in X^*$ attains it norm on an element $u, \quad ||u|| = 1$.

f) X^* is uniformly convex if and only if the norm $||\cdot||_X$ is uniformly differentiable on the set $\partial B_1 := \{u : ||u|| = 1,\ u \in X\}$.

g) $||\cdot||_X$ is Gâteaux-differentiable on $X\backslash\{0\}$ if and only if X^* is strictly convex.

Proofs of Theorem 16.7.1 and of many other results in this Section are omitted. One can find them in [De].

Definition 16.7.2. *The map*

$$j : X \to 2^{X^*}, \quad ju = \{u^* \in X^* : ||u^*|| = ||u||, \quad u^*(u) = ||u||^2\} \tag{16.7.1}$$

is called the duality map .

Theorem 16.7.2. *a) The set $\{ju\}$ is convex, $j(\lambda u) = \lambda j(u) \quad \forall \lambda \mathbb{R}$.*

b) The set $\{ju\}$ consists of one element if and only if X^* is strictly convex; $j = I$, the identity operator, if $X = H$, the Hilbert space.

Semi-inner products in a Banach space are defined by the formulas:

$$||v|| \lim_{t\to 0, t>0} t^{-1}(||v + tu|| - ||v||) := (u, v)_+. \tag{16.7.2}$$

$$||v|| \lim_{t\to 0, t>0} t^{-1}(||v|| - ||v - tu||) := (u, v)_-. \tag{16.7.3}$$

The existence of the limit in (16.7.2) follows from the properties of the norm:

$$\begin{aligned} ||v + su|| - ||v|| &= ||\frac{s}{t}v + \frac{s}{t}tu + \left(1 - \frac{s}{t}\right)v|| - ||v|| \\ &\le \frac{s}{t}(||v + tu|| - ||v||), \quad 0 < s < t. \end{aligned} \tag{16.7.4}$$

Thus the function

$$\frac{||v + su|| - ||v||}{s} \tag{16.7.5}$$

is nondecreasing. This function is bounded from below as $s \to 0, \quad s > 0$:

$$\frac{||v + su|| - ||v||}{s} \ge -||u||. \tag{16.7.6}$$

This inequality follows from the triangle inequality:

$$||v + su|| \ge |||v|| - s||u||| \ge ||v|| - s||u||. \tag{16.7.7}$$

Therefore the limits in (16.7.2) and (16.7.3) exist.

Theorem 16.7.3. *One has*

$$(u,v)_+ = (u,v)_- \tag{16.7.8}$$

if and only if X^* is strictly convex.

If $u(t) \in C^1(a,b;X)$ then $||u(t)|| := g(t)$ satisfies the equations:

$$g(t)D_+g = (\dot{u}(t), u(t))_+, \quad g(t)D_-g = (\dot{u}(t), u(t))_-, \tag{16.7.9}$$

where

$$D_+g = \overline{\lim_{h\to 0,\ h>0}} \frac{g(t+h)-g(t)}{h}; \quad D_-g = \overline{\lim_{h\to 0,\ h>0}} \frac{g(t)-g(t-h)}{h}. \tag{16.7.10}$$

Definition 16.7.3. *An operator $F : D \subset X \to X$ is called accretive if*

$$(F(u)-F(v), u-v)_+ \geq 0 \quad \forall u,v \in D, \tag{16.7.11}$$

strictly accretive if

$$(F(u)-F(v), u-v)_+ > 0 \quad \forall u,v \in D, \quad u \neq v, \tag{16.7.12}$$

strongly accretive if

$$(F(u)-F(v), u-v)_+ \geq c||u-v||^2 \quad \forall u,v \in D, \tag{16.7.13}$$

maximal accretive (or m-accretive) if

$$(F(u)-f, u-v)_+ \geq 0 \quad \forall u \in D \tag{16.7.14}$$

implies $v \in D$ and $F(v) = f$,

and hyperaccretive if (16.7.11) holds and

$$R(F+\lambda I) = X, \tag{16.7.15}$$

for some $\lambda = const > 0$.

Theorem 16.7.4. *Assume that $F : D \to X$ is hyperaccretive. Then $R_\lambda := (I+\lambda F)^{-1}$, $\lambda > 0$, is nonexpansive, $\lim_{\lambda\to 0} R_\lambda(u) = u$, $||R_\lambda(u)|| \leq ||F(u)||$. If X and X^* are uniformly convex then $\lim_{\lambda\to 0} ||FR_\lambda(u) - F(u)|| = 0$ and the problem*

$$\dot{u} = -F(u), \quad u(0) = u_0, \quad u_0 \in D, \tag{16.7.16}$$

has a unique solution on $[0,\infty)$. The solution $u(t)$ is continuous, weakly differentiable, and $||\dot{u}(t)||$ is decreasing.

Consider the operator

$$U(t)u_0 = u(t; u_0), \tag{16.7.17}$$

where $u(t; u_0)$ solves (16.7.16). We have

$$U(0) = I, \quad \lim_{t\to 0,\ t>0} ||U(t)u_0 - u_0|| = 0, \tag{16.7.18}$$

$$U(t+s) = U(t)U(s). \tag{16.7.19}$$

The family $\{U(t)\}$ forms a semigroup. In [Mi] one finds a presentation of the nonlinear semigroup theory, and in [P] the linear semigroup theory is presented.

16.8 Existence of solutions to operator equations

In this Section we mention briefly the methods for proving the existence results for solutions to operator equations. Much more the reader finds in [Br, GG, KA, Kr, KZ, KV, Z].

For linear equations

$$u = Au + f \tag{16.8.1}$$

in a Banach space X, assuming that A is a linear operator, one has the existence of a unique solution for any f provided that $||A|| < 1$. This follows from the contraction mapping principle. The solution can be calculated by the iterative method

$$u_{n+1} = Au_n + f, \quad u_0 = u_0, \tag{16.8.2}$$

where $u_0 \in X$ is arbitrary. The method converges at the rate of geometrical series with the denominator $q = ||A|| < 1$.

If A is compact, then the Fredholm-Riesz theory applies, (see e.g., [Y]). This theory yields the existence of a unique solution to (16.8.1) for any $f \in X$ provided that $\mathcal{N}(I-A) = \{0\}$, where $\mathcal{N}(I-A)$ is the null-space of the operator $I - A$. If $\mathcal{N}(I-A) \neq \{0\}$, then equation (16.8.1) is solvable if and only if $< u^*, f > \ > 0$ for all $u^* \in \mathcal{N}(I - A^*)$, where A^* is the adjoint to A operator. This is the well known Fredholm alternative .

If A is nonlinear, then the existence of a solution to the equation

$$u = A(u) \tag{16.8.3}$$

is often proved by applying topological methods, such as degree theory. Typical examples are Schauder's principle , Rothe's theorem, and Leray-Schauder's principle.

Theorem 16.8.1. *(Schauder) If a continuous in X operator A maps a convex closed set D into its compact subset, then equation (16.8.3) has a solution in D.*

Theorem 16.8.2. *(Rothe) If A is compact on $\bar{D}$, where $D \subset X$ is a bounded convex domain, and $A(\partial D) \subset \bar{D}$, where ∂D is the boundary of D, then equation (16.8.2) has a solution in $\bar{D}$.*

Let $A(\cdot, \lambda)$ be a parametric family of compact operators, $0 \leq \lambda \leq 1$, $A(u, 1) = A(u)$, and equation $u = A(u, 0)$ has a solution,

$$||A(u, 0)|| \leq b \quad \forall u \in \{u : ||u|| = b\}. \tag{16.8.4}$$

Theorem 16.8.3. *(Leray-Schauder). Under the above assumptions equation (16.8.3) has a solution provided that*

$$||u(\lambda)|| \leq a < b \qquad \forall \lambda \in [0, 1], \tag{16.8.5}$$

where $\{u(\lambda)\}$ is the set of all solutions to the equation

$$u = A(u, \lambda), \quad \lambda \in [0, 1]. \tag{16.8.6}$$

Theorem 16.8.3 is often used in the following form:
If all the solutions $u(\lambda)$ to the equation

$$u = \lambda A(u), \quad \lambda \in [0, 1] \tag{16.8.7}$$

satisfy the a priori estimate

$$||u(\lambda)|| \leq a, \tag{16.8.8}$$

then equation (16.8.3) has a solution in the ball $\{u : ||u|| \leq a\}$.

Definition 16.8.1. *A map A in a Banach space X is called a generalized contraction mapping if*

$$||A(u) - A(v)|| \leq q(a, b)||u - v||, \quad 0 < a \leq ||u - v|| \leq b, \tag{16.8.9}$$

where $q(a, b) < 1$.

For example, A is a generalized contraction mapping if

$$||A(u) - A(v)|| \leq ||u - v|| - g(||u - v||), \tag{16.8.10}$$

where $g(s) > 0$ if $s > 0$, $g(0) = 0$, and g is a continuous function on $[0, \infty)$.

Theorem 16.8.4. *(Krasnoselsky) If A is a generalized contraction mapping on a closed set D and $AD \subset D$, then equation (16.8.3) has a solution in D.*

Proof of Theorem 16.8.4. The proof follows [KV]. Consider the sequence

$$u_{n+1} = A(u_n), \quad u_0 \in D. \tag{16.8.11}$$

We have

$$\begin{aligned} ||u_{n+2} - u_{n+1}|| &= ||A(u_{n+1} - A(u_n)|| \\ &\leq q(a,b)||u_{n+1} - u_n|| < ||u_{n+1} - u_n||. \end{aligned} \tag{16.8.12}$$

Therefore there exists

$$\lim_{n\to\infty} ||u_{n+1} - u_n|| = s < \infty. \tag{16.8.13}$$

We claim that $s = 0$. Indeed, if $s > 0$ then

$$s \leq ||u_{n+1} - u_n|| \leq s + \epsilon, \qquad \forall n > n(\epsilon), \tag{16.8.14}$$

and

$$s \leq ||u_{n+m} - u_{n+m-1}|| \leq q^m(s, s+\epsilon)(s+\epsilon), \quad \forall n > n(\epsilon). \tag{16.8.15}$$

If m is sufficiently large, then $q^m(s, s+\epsilon)$ is as small as one wishes, so (16.8.15) yields a contradiction which proves that $s = 0$.

Let us prove that

$$AB(u_p, \epsilon) \subset B(u_p, \epsilon), \tag{16.8.16}$$

where $\epsilon > 0$ is an arbitrary small fixed number, and u_p is chosen so that

$$d_p := ||u_{p+1} - u_p|| \leq \frac{\epsilon}{2}\left[1 - q\left(\frac{\epsilon}{2}, \epsilon\right)\right]. \tag{16.8.17}$$

Such a number p exists because

$$\lim_{n\to\infty} d_n := \lim_{n\to\infty} ||u_{n+1} - u_n|| = 0, \tag{16.8.18}$$

as we have proved.

If (16.8.16) is verified then the sequence $\{u_n\}$ is a Cauchy sequence, so it has a limit

$$\lim_{n\to\infty} u_n = u. \tag{16.8.19}$$

The limit u solves equation (16.8.3) as one can see by passing to the limit $n \to \infty$ in equation (16.8.11). Uniqueness of the solution (16.8.3) follows from (16.8.9). Thus, the proof is completed if (16.8.16) is verified.

Let us verify (16.8.16) . Let $||u - u_p|| \leq \frac{\epsilon}{2}$. Then

$$\begin{aligned} ||A(u) - u_p|| &\leq ||A(u) - A(u_p)|| + ||A(u_p) - u_p|| \\ &\leq ||u - u_p|| + d_p \leq \frac{\epsilon}{2} + \frac{\epsilon}{2} = \epsilon. \end{aligned} \tag{16.8.20}$$

Let $\frac{\epsilon}{2} < ||u - u_p|| \leq \epsilon$. Then, using (16.8.17), one gets

$$\begin{aligned} ||A(u) - u_p|| &\leq ||u - u_p|| + d_p \leq q\left(\frac{\epsilon}{2}, \epsilon\right)\epsilon + d_p \\ &\leq q\left(\frac{\epsilon}{2}, \epsilon\right)\epsilon + \frac{\epsilon}{2} - \frac{\epsilon}{2}q\left(\frac{\epsilon}{2}, \epsilon\right) \leq \epsilon \end{aligned} \tag{16.8.21}$$

Thus, (16.8.16) is verified, and Theorem 16.8.4 is proved. □

Theorem 16.8.4 generalizes Theorem 16.1.1, because if condition (16.1.1) holds, then condition (16.8.9) holds.

Theorem 16.8.5. *(Krasnoselsky). Let $A : D \to D$ where D is a convex bounded, closed subset of a Banach space X. Assume that $A = B + T$, where B is a generalized contraction map and T is compact. Then equation (16.8.3) has a solution $u \in D$.*

Definition 16.8.2. *A map $A : X \to X$ is called nonexpansive if*

$$||A(u) - A(v)|| \leq ||u - v||. \tag{16.8.22}$$

Let H denote a Hilbert space.

Theorem 16.8.6. *If $A : D \to D$ is nonexpansive, $D \subset H$ is a bounded, convex, and closed set, then equation (16.8.3) has a solution $u \in D$.*

Definition 16.8.3. *A closed convex set $K \subset X$ is called a cone if $u \in K$ and $u \neq 0$ imply that $\lambda u \in K \quad \forall \lambda \geq 0$ and $\lambda u \notin K \quad \forall \lambda \in \mathbb{R}$.*

We write $u \geq v$ if $u - v \in K$. Elements of K are called positive elements. A cone is called solid if there is an element $u \in K$ such that $B(u, r) \subset K$ for some $r > 0$, where $B(u, r) := \{v : ||u - v|| \leq r\}$. A cone is called reproducing if every element $w \in X$ can be represented as $w = u - v$, where $u, v \in K$. A cone is called normal iff $0 \leq u \leq v$ implies $||u|| \leq N||v||$, where the constant N does not depend on u and v.

Definition 16.8.4. *An operator $A : X \to X$ in a Banach space X with a cone K is called K-monotone if $u \leq v$ implies $A(u) \leq A(v)$.*

If A is K-monotone and $u \leq v$, $A(u) \geq u$, $A(v) \leq v$, then $u \leq w \leq v$ implies $u \leq A(w) \leq v$.

A cone is called strongly minihedral if for any bounded set $U = \{u\} \in K$ has a supremum, that is an element which is the minimal element in the set $W = \{w\} \in K$ of the elements such that $u \leq w$, $\forall u \in U$.

Let $[u_1, u_2] := \{u : u_1 \leq u \leq u_2, \ u \in K\}$.

Theorem 16.8.7. *If K is strongly minihedral, and $A[u_1, u_2] \subset [u_1, u_2]$. Then equation (16.8.3) has a solution $u \in [u_1, u_2]$.*

Proof. The set W of elements $w \in [_1, u_2]$ such that $A(w) \geq w$ is non-void: it contains u_1. Also, $AW \subset W$. Let $s = \sup W$. Then $w \in W$ implies $w \leq A(w) \leq A(s)$, so $s \leq A(s)$ and $s \in W$. Therefore $A(s) \in W$. Consequently $A(s) \leq s$. Thus, $A(s) = s$.

Theorem 16.8.7 is proved. □

Example 16.8.1. Consider the problem

$$\dot{u} = f(t, u), \quad u(0) = u_0 \tag{16.8.23}$$

in a Banach space X. Assume that

$$f(t, u) = g(t, u) + h(t, u) \tag{16.8.24}$$

where

$$||g(t, u_1) - g(t, u_2)|| \leq L||u_1 - u_2||, \quad L = \text{const} > 0, \tag{16.8.25}$$

and $h(t, \cdot)$ is a compact operator. Problem (16.8.23) can be written as

$$u(t) = u_0 + \int_0^t f(s, u(s))ds = Bu + Tu, \tag{16.8.26}$$

where

$$\begin{aligned} Bu &:= u_0 + \int_0^t g(s, u(s))ds, \\ Tu &:= \int_0^t h(s, u(s))ds. \end{aligned} \tag{16.8.27}$$

Consider equation (16.8.26) in the space $C([0, \delta], X)$ of continuous on $[0, \delta]$ functions $u(t)$ with values in X. Then the operator B is a contraction mapping in this space if $\delta > 0$ is sufficiently small, and T is compact in this space. By Theorem 16.8.5 problem (16.8.26) has a solution in $C([0, \delta], X)$ for sufficiently small $\delta > 0$.

16.9 Compactness of embeddings

The basic result of this Section is:

Theorem 16.9.1. *Let $X_1 \subset X_2 \subset X_3$ be Banach spaces,*

$$||u||_1 \geq ||u||_2 \geq ||u||_3,$$

that is, the norms are comparable, and if $||u_n||_3 \to 0$ as $n \to \infty$ and u_n is fundamental in X_2, then $||u_n||_2 \to 0$, (i.e., the norms in X_2 and X_3 are compatible). Under the above assumptions the embedding operator $i : X_1 \to X_2$ is compact if and only if the following two conditions are valid:

a) The embedding operator $j : X_1 \to X_3$ is compact,

and

b) The following inequality holds:

$$||u||_2 \leq s||u||_1 + c(s)||u||_3, \quad \forall u \in X_1, \quad \forall s \in (0,1),$$

where $c(s) > 0$ is a constant.

This result is an improvement of the author's old result [R1]. We follow [R61]. We construct a counterexample to a theorem in [B], p.35, where the validity of the inequality b) in Theorem 16.9.1 is claimed without the assumption of the compatibility of the norms of X_2 and X_3, see Remark 16.9.1 at the end of this Section.

Proof of Theorem 16.9.1.

1. *The sufficiency of conditions a) and b) for compactness of $i : X_1 \to X_2$.*

Assume that a) and b) hold, and let us prove the compactness of i. Let $S = \{u : u \in X_1, ||u||_1 = 1\}$ be the unit sphere in X_1. Using assumption a), select a sequence u_n which converges in X_3. We claim that this sequence converges also in X_2. Indeed, since $||u_n||_1 = 1$, one uses assumption b) to get

$$||u_n - u_m||_2 \leq s||u_n - u_m||_1 + c(s)||u_n - u_m||_3 \leq 2s + c(s)||u_n - u_m||_3.$$

Let $\eta > 0$ be an arbitrary small given number. Choose $s > 0$ such that

$$2s < \frac{1}{2}\eta,$$

and for a fixed s choose n and m so large that

$$c(s)||u_n - u_m||_3 < \frac{1}{2}\eta.$$

This is possible because the sequence u_n converges in X_3. Consequently,

$$||u_n - u_m||_2 \leq \eta$$

if n and m are sufficiently large. This means that the sequence u_n converges in X_2. Thus, the embedding $i : X_1 \to X_2$ is compact. In the above argument, i.e., in the proof of the sufficiency, the compatibility of the norms was not used.

2. *The necessity of the compactness of $i : X_1 \to X_2$ for conditions a) and b) to hold.*

Assume now that i is compact. Let us prove that conditions a) and b) hold. In the proof of the necessity of these conditions the assumption about the compatibility of the norms of X_2 and X_3 is used essentially. Without this assumption one cannot prove that conditions a) and b) hold. This is demonstrated in Remark 16.9.1 after the end of the proof of the Theorem.

If i is compact, then assumption a) holds because $||u||_2 \geq ||u||_3$. Suppose that assumption b) fails. Then there is a sequence u_n and a number $s_0 > 0$ such that $||u_n||_1 = 1$ and

$$||u_n||_2 \geq s_0 + n||u_n||_3. \tag{16.9.1}$$

If the embedding operator i is compact and $||u_n||_1 = 1$, then one may assume that the sequence u_n converges in X_2. Its limit cannot be equal to zero, because, by (1),

$$||u_n||_2 \geq s_0 > 0.$$

The sequence u_n converges in X_3 because of the inequality

$$||u_n - u_m||_2 \geq ||u_n - u_m||_3$$

and because the sequence u_n converges in X_2.

Its limit in X_3 is not zero, because the norms in X_3 and in X_2 are compatible.

Thus,

$$\lim_{n\to\infty} ||u_n||_3 > 0.$$

Consequently, inequality (1) implies

$$||u_n||_3 = O(\frac{1}{n}) \to 0 \quad \text{as} \quad n \to \infty,$$

while

$$\lim_{n\to\infty} ||u_n||_3 > 0.$$

This is a contradiction, which proves that b) holds.

Theorem 16.9.1 is proved. □

Remark 16.9.1. In [B], p. 35, the following claim is stated:

Claim. Let $X_1 \subset X_2 \subset X_3$ be three Banach spaces. Suppose the embedding $X_1 \to X_2$ is compact. Then given any $\epsilon > 0$, there is a $K(\epsilon) > 0$, such that

$$||u||_2 \leq \epsilon ||u||_1 + K(\epsilon)||u||_3$$

for all $u \in X_1$.

This claim is *not correct* because there is no assumption about the compatibility of the norms of X_2 and X_3.

For example, let $L^2(0,1)$ be the usual Lebesgue space of square integrable functions, $X_3 = L^2(0,1)$, and X_2 be a Banach space of $L^2(0,1)$ functions with a finite value at a fixed point $y \in [0,1]$ and with the norm

$$||u||_2 := ||u||_{L^2(0,1)} + |u(y)| = ||u||_3 + |u(y)|.$$

The space X_2 is complete because X_3 is complete and the one-dimensional space, consisting of numbers $u(y)$ with the usual norm $|u(y)|$, is complete. A function $u_0(x) = 0$ for $x \neq 0$ and $u_0(y) = 1$ has the properties

$$||u_0||_3 = 0, \quad ||u_0||_2 = 1.$$

One has $X_2 \subset X_3$. The norms in X_2 and X_3 are *comparable*, i.e., $||u||_3 \leq ||u||_2$. However, these norms are *not compatible*: there is a convergent to zero sequence $\lim_{n\to\infty} u_n = 0$ in X_3 such that it does not converge to zero in X_2, for example, $\lim_{n\to\infty} ||u_n||_2 = 1$ in X_2. For instance, one may take $u_n(x) = u_0(x)$ for all $n = 1, 2, \dots,$ and an arbitrary fixed $y \in [0,1]$. Then $||u_n||_2 = 1$ and $||u_n||_3 = 0$, $\lim_{n\to\infty} ||u_n||_2 = 1$ and $\lim_{n\to\infty} ||u_n||_3 = 0$. The sequence u_n converges to zero in X_3 and to a non-zero element u_0 in X_2. In this case inequality (16.9.1) holds for any fixed $s_0 \in (0,1)$ and any n, but the contradiction, which was used in the proof of the necessity in Theorem 16.9.1, can not be obtained because $||u_n||_3 = 0$ for all n.

Let us construct a counterexample which shows that the Claim, mentioned above, is not correct. Fix a $y \in [0,1]$. Choose the one-dimensional space of functions $\{u : u = \lambda u_0(x)\}$ as X_1, where $\lambda = const$ and $u_0(x)$ was defined above, and define the norm in X_1 by the formula $||u||_1 = |\lambda|$. Let $X_3 = L^2(0,1)$. The space X_1 is a one-dimensional Banach space. Therefore bounded sets in X_1 are precompact. Note that $|\lambda| = ||\lambda u_0||_1 = ||\lambda u_0||_2 \geq ||\lambda u_0||_3 = 0$ because $||u_0||_3 = 0$. Here the Banach space X_2 is defined as above with the norm $||u||_2 := ||u||_{L^2(0,1)} + |u(y)|$, and the equalities $||u_0||_2 = 1$ and $||u_0||_3 = 0$ are used.

Consequently,

$$X_1 \subset X_2 \subset X_3, \quad ||u||_1 \geq ||u||_2 \geq ||u||_3,$$

and the embedding $i : X_1 \to X_2$ is compact because bounded sets in finite-dimensional spaces are precompact and X_1 is a one-dimensional space. Thus, all the assumptions of the Claim are satisfied. However the inequality of the Claim:

$$||u||_2 \leq \epsilon ||u||_1 + K(\epsilon)||u||_3 \quad \forall u \in X_1$$

does not hold for any fixed $\epsilon \in (0, 1)$. In our counterexample

$$u = \lambda u_0, \quad ||u_0||_3 = 0,$$

and the above inequality takes the form:

$$|\lambda| \leq \epsilon |\lambda|.$$

Clearly, this inequality does not hold for a fixed $\epsilon \in (0, 1)$ unless $\lambda = 0$.

Bibliographical notes

The contents of this book is based on the author's papers cited in the bibliography. The most important of the preceding papers was the paper by M. Gavurin [Ga], who deals with a continous analog of Newton's method. There is a large literature on solving ill-posed problems (see e.g. [BG, I, M, R44, VV, VA, TLY], to mention a few). The DSM as a tool for solving operator equations, especially ill-posed and possibly nonlinear, is developed in this book systematically.

The examples of inverse and ill-posed problems, mentioned in Section 2.1, are partly taken from [R19] and [R44]. Some of these examples are discussed in more detail in many books and papers. We mention the books [AV, MY] and the papers [R2, R3, R5, R6] on antenna synthesis. Variational regularization (Section 2.2) is discussed in the books [BG, I, M, R44, TLY, VV, VA], to mention a few. Our presentation contains several new points. We deal with unbounded, densely defined, closed linear operators and define the operator $(A^*A_aI)^{-1}A^*$ for $a = \text{const} > 0$ on the whole Hilbert space in the case when the domain $D(A^*)$ is dense in H, but is not the whole space. This allows us to extend to the case of unbounded operators the usual theory of variational regularization without requiring compactness properties from the stabilizing functional [R58, R59, R62]. We give a new discrepancy principle which does not require to solve theoretically exactly the usual discrepancy principle equation, but rather to find an approximate minimizer of the functional which is used in the standard theory of variational regularization. (Theorem 2.2.5) (see [R44, R48]). We formulate a new notion of regularizer ([R35]).

Section 2.3, Quasisolutions, contains some well-known material (see, e.g., [I]).

Section 2.4, Iterative regularization, contains a proof of the following general result: every solvable linear equation in a Hilbert space is solvable by a convergent iterative process (Theorem 2.4.1). There are many papers and books on iterative methods (see, e.g., [BG, VA, VV]).

Section 2.5, Quasiinversion, presents very briefly the idea of the quasiinversion method for solving ill-posed problems. A simple example of the application of this method is given in Section 2.5. More material on the quasiinversion methods one finds in the book [LL].

In Section 2.6 the idea of the dynamical systems method (DSM) is discussed.

In Section 2.7 variational regularization for nonlinear operator equations is discussed (cf [R33, R48]). Nonlinear ill-posed problems are discussed in [TLY].

Section 3.1 follows [R40]. The results presented in this Section generalize some results from [AR]. Example 3.1.1 was not published earlier. Sections 3.2 - 3.7 contain applications of the results obtained in Section 3.1. Various versions of the DSM are constructed in these Sections for continuous analogs of classical methods, including Newton's method, modified Newton's method, Gauss-Newton's method, gradient method, simple iteration method, and a minimization method.

The results of Chapter 4 are based on the papers [R29, R60, R62]. Section 4.4 uses some ideas from [R10].

In Section 4.5 a new approach is fiven to stable calculation of values of unbounded operators. This problem has been ealier treated by the variational regularization method in [I, M].

In Chapter 5 some auxiliary inequalities are presented. Theorems 5.1.1 and 5.2.1 are used in the following Chapters. The first version of Theorem 5.1.1 appeared first in [ARS] and its refinement was given in [AR], [R37, R44].

An erroneous version of Theorem 5.3.1 appeared in [Al]. A counterexample to the claim in [Al] was given in [R44], where a corrected version of the theorem has been proved and the basic idea of the proof from[Al] was used (see also [ARS]). Theorem 5.3.2 appeared in [ZS]. Our proof is shorter. The results in Section 5.4 can be found, for example, in [Te].

The results of Section 6.1, Chapter 6, are known, but our presentation is self-contained. The main results of Chapter 6 are given in Sections 6.2 and 6.3. Theorem 6.2.1 is taken from [R37]. Its earlier versions appeared in [ARS] and [AR], see also [R44].

The results of Section 6.3 are taken from [R44].

Sections 7.1 and 7.2 of Chapter 7 are based on the results from [R44, R63], and Section 7.3, Theorem 7.3.1, is taken from [R63, R43].

Chapter 8 is based on the papers [R49, R56].

Chapter 9 is based on papers [R52, R54], see also [R44, R49].

Chapter 10 is based on the results from [R44] in the case of well-posed problems, and on [R32] in the case of ill-posed problems.

Chapter 11 is based on [R55].

Chapter 12 is based on [R38, R44].

Chapter 13 is based on [R39, R42, R53].

There is a long history of the results preceding Theorem 13.2.2 which gives a sufficient condition for a local homeomorphism to be a global one. It starts with the Hadamard's paper of 1906 [Ha], and its further developments are described in [OR]. Our approach is purely analytical. There are also approaches based on algebraic topology ([Sp]).

Sections 14.2 of Chapter 14 is based on [R44]. Originally these results appeared in [AR]. Our presentation is slightly different. Section 14.3 is based on [R43] and Section 14.4 is based on [R63].

Section 15.1 is based on [R4], where for the first time the idea of using the stepsize as a regularizing parameter was proposed and implemented in a solution of the stable numerical differentiation of noisy data. This idea has been used in many papers of various authors and in many applications (see also [R7, R10, R11, R12], Appendix 4 in [R13], [R28], Section 7.3 in [R50], [R30, R34].

Section 15.2 is based on [R45, R57]. It follows closely [R57].

Section 15.3 follows closely [R8].

Section 15.4 was not published earlier in this form, but it is based on the known ideas. Section 15.5 is based on [R30] and [ARU].

Section 15.6 follows closely [R18].

Chapter 16 contains auxiliary material. The material from Sections 16.1-16.5 is well known and can be found, for example, in [De].

The material from Sections 16.7 and 16.8 can be found in the books [De], [Kr], [KZ], and in other books, cited in these Sections.

Section 16.6 is based on [R41].

Section 16.9 is based on [R1] and follows closely [R61].

Bibliography

[ARU] Ahn Soyoung, A. G. Ramm, U Jin Choi, A scheme for a stable numerical differentiation, *Jour. Comp. Appl. Math*, 186, N2, (2006), 325-334.

[AR] R. Airapetyan, A. G. Ramm, Dynamical systems and discrete methods for solving nonlinear ill-posed problems, *Appl.Math.Reviews*, vol. 1, Ed. G. Anastassiou, World Sci. Publishers, 2000, pp.491-536.

[ARS] R.Airapetyan, A.Smirnova, A.G.Ramm, Continuous methods for solving nonlinear ill-posed problems, in the book "*Operator theory and applications*", Amer. Math. Soc., Fields Institute Communications, Providence,RI, 2000, pp. 111-138.

[A] N. Akhiezer, *Lectures on approximation theory*, Nauka, Moscow, 1965.

[Al] Y. Alber, A new approach to the investigation of evolution differential equations in Banach spaces, *Nonlin. Analysis, Theory, Methods and Applic.*, 23, N9, (1994), 1115-1134.

[AV] M. Andrijchuk, N. Voitovich, D. Savenko, V. Tkachuk, *Synthesis of antenna*, Naukova Dumka, Kiev, 1993.

[BG] A. Bakushinsky, A. Goncharsky, *Iterative methods for solving ill-posed problems*, Nauka, Moscow, 1989 (Russian).

[BL] J. Bergh, J, Löfstrom, *Interpolation spaces*, Springer, New York, 1976.

[B] M. Berger, *Nonlinearity and functional analysis*, Acad. Press, New York, 1977.

[Br] F. Browder, *Nonlinear operators and nonlinear equations of evolution in Banach spaces*, Amer. Math. Soc., Providence, RI, 1976.

[D] I. Daubechies, *Ten lectures on wavelets*, SIAM, Philadelphia, 1992.

[DR] P. Davis, P. Rabinowitz, *Methods of numerical integration*, Acad. Press, New York, 1989.

[De] K. Deimling, *Nonlinear functional analysis*, Springer, New York, 1985.

[DS] N. Dunford, J. Schwartz, *Linear operators, Interscience*, New York, 1958.

[GG] H. Gajewski, K. Gröger, K. Zacharias, *Nichtlineare operatorgleichungen und operator differentialgleichungen*, Akad Verlag, Berlin, 1974.

[GW] A.Galperin, Z.Waksman, *Ulm's method under regular smoothness*, Num. Funct. Anal. Optim., 19, (1998), 285-307.

[Ga] M. Gavurin, Nonlinear functional equations and continuous analysis of iterative methods, *Izvestiya Vusov, Math.*, 5, (1958), 18-31 (Russian).

[G] I. Glazman, *Direct methods of qualitative spectral analysis of differential operators*, Moscow, Nauka, 1963.

[GR] V. Gol'dshtein, A. G. Ramm, Embedding operators for rough domains, *Math. Ineq. and Applic.*, 4, N1, (2001), 127-141.

[GR1] V. Gol'dshtein, A. G. Ramm, Embedding operators and boundary-value problems for rough domains, *Intern. Jour. of Appl. Math. Mech.*,1, (2005), 51-72.

[Ha] J. Hadamard, Sur les transformations ponctuelles, *Bull. Soc. Math. France*, 34, (1906), 71-84.

[HW] G. Hall, J. Watt (editors), *Modern numerical methods for ordinary differential equations*, Clarendon Press, Oxford, 1976.

[H] P. Hartman, *Ordinary differential equations*, Wiley, New York, 1964.

[I] V. Ivanov, V. Tanana, V. Vasin, *Theory of ill-posed problems and its applications*, VSP, Utrecht, 2002.

[J] J. Jerome, *Approximation of nonlinear evolution systems*, Acad. Press, New York, 1983.

[KNR] B.Kaltenbacher, A.Neubauer, A.G. Ramm, Convergence rates of the continuous regularized Gauss-Newton method, *Jour. Inv. Ill-Posed Probl.*, 10, N3, (2002), 261-280.

[Ka] E. Kamke, *Differential Gleichungen. Lösungsmethoden und Lösungen*, Akad. Verlag., Leipzig, 1959.

[KA] L. Kantorovich, G. Akilov, *Functional analysis*, Pergamon press, New York, 1982.

[K] T. Kato, *Perturbation theory for linear operators*, Springer Verlag, New York, 1984.

[KR1] A. I. Katsevich, A. G. Ramm, Multidimensional algorithm for finding discontinuities of functions from noisy data. *Math. Comp. Modelling*, 18, N1, (1993), 89-108.

[KR2] A. I. Katsevich, A. G. Ramm, Nonparametric estimation of the singularities of a signal from noisy measurements, *Proc. AMS*, 120, N8, (1994), 1121-1134.

[Kr] M. Krasnoselsky, *Positive solutions of operator equations*, Groningen, Noordhoff, 1964.

[KZ] M. Krasnoselsky, P. Zabreiko, *Geometrical methods of nonlinear analysis*, Springer, New York, 1984.

[KV] M. Krasnoselsky *et al*, *Approximate solutions of operator equations*, Groningen, Noordhoff, 1972.

[LL] J. Lattes, J. Lions, *Méthode de quasi-réversibilité et applications*, Dunod, Paris, 1967.

[L] J. Lions, *Quelques méthodes de résolution des problemes aux limites nonlineatres*, Dunod, Paris, 1969.

[Li] O. Liskovetz, Regularization of equations with a closed linear operator, *Diff. equations*, 7, (1970), 972-976 (Russian).

[MY] B. Minkovich, V. Yakovlev, *Theory of antenna synthesis*, Sov. Radio, Moscow, 1969.

[Mi] I. Miyadera, *Nonlinear semigroups*, Amer. Math. Soc., Providence, RI, 1977.

[M] V. Morozov, *Methods of solving incorrectly posed problems*, Springer Verlag, New York, 1984.

[N] M. Naimark, *Linear differential operators*, Ungar, New York, 1969.

[OR] J. Ortega, W. Rheinboldt, *Iterative solution of nonlinear equations in several variables*, SIAM, Philadelphia, 2000.

[P] A. Pazy, *Semigroups of linear operators and applications to partial differential equations*, Springer, New York, 1983.

[PS] G. Polya, G. Szego, *Problems and theorems in analysis*, Springer Verlag, New York, 1983.

[Pow] M. Powell, *Approximation theory and methods*, Cambridge Univ. Press, Cambridge 1981.

[R1] A. G. Ramm, A necessary and sufficient condition for compactness of embedding. *Vestnik Lenigr. Univ. (Vestnik)* N 1, (1963), 150-151.

[R2] A. G. Ramm, Antenna synthesis with the prescribed pattern, *22 sci. session dedicated the day of radio*, Moscow, 1966, Section of antennas, 9-13.

[R3] A. G. Ramm, Optimal solution of the antenna synthesis problem, *Doklady Acad. of Sci.* USSR, 180, (1968), 1071-1074.

[R4] A. G. Ramm, On numerical differentiation, *Mathem., Izvestija vuzov*, 11, (1968), 131-135.

[R5] A. G. Ramm, Nonlinear antenna synthesis problems, *Doklady Acad. Sci.* USSR, 186, (1969), 1277-1280.

[R6] A. G. Ramm, Optimal solution of the linear antenna synthesis problem, *Radiofisika*, 12, (1969), 1842-1848. 43 #8223.

[R7] A. G. Ramm, Simplified optimal differentiators, *Radiotech.i Electron.* 17, (1972), 1325-1328.

[R8] A. G. Ramm, On simultaneous approximation of a function and its derivative by interpolation polynomials, *Bull. Lond. Math. Soc.* 9, (1977), 283-288.

[R9] A. G. Ramm, Stationary regimes in passive nonlinear networks, in "*Nonlinear Electromagnetics*", Ed. P.L.E. Uslenghi, Acad. Press, N. Y., 1980, pp. 263-302.

[R10] A. G. Ramm, Stable solutions of some ill-posed problems, *Math. Meth. in the appl. Sci.* 3, (1981), 336-363.

[R11] A. G. Ramm, Estimates of the derivatives of random functions. *J. Math. Anal. Appl.*, 102, (1984), 244-250.

[R12] A. G. Ramm, T. Miller, Estimates of the derivatives of random functions II, *J. Math. Anal. Appl.* 110, (1985), 429-435.

[R13] A. G. Ramm, *Scattering by obstacles*, D. Reidel, Dordrecht, 1986, pp.1-442.

[R14] A. G. Ramm, Characterization of the scattering data in multidimensional inverse scattering problem, in the book: "*Inverse Problems: An Interdisciplinary Study.*" Acad. Press, NY, 1987, 153-167. (Ed. P. Sabatier).

[R15] A. G. Ramm, Necessary and sufficient conditions for a function to be the scattering amplitude corresponding to a reflecting obstacle, *Inverse problems*, 3, (1987), L53-57.

[R16] A. G. Ramm, Completeness of the products of solutions to PDE and uniqueness theorems of inverse scattering, *Inverse problems*, 3, (1987), L77-L82

[R17] A. G. Ramm, Recovery of the potential from fixed energy scattering data. *Inverse problems*, 4, (1988), 877-886; 5, (1989) 255.

[R18] A. G. Ramm, A. van der Sluis, Calculating singular integrals as an ill-posed problem, *Numer. Math.*, 57, (1990) 139-145.

[R19] A. G. Ramm, *Multidimensional inverse scattering problems*, Longman/Wiley, New York, 1992, pp.1-385.

[R21] A. G. Ramm, A. Zaslavsky, Reconstructing singularities of a function given its Radon transform, *Math. Comp.Modelling*, 18, N1, (1993), 109-138.

[R23] A. G. Ramm, Optimal local tomography formulas, *Pan Amer. Math. Journ.*, 4, N4, (1994), 125-127.

[R24] A. G. Ramm, Finding discontinuities from tomographic data, *Proc. Amer. Math. Soc.*, 123, N8, (1995), 2499-2505.

[R25] A. G. Ramm, A. I. Katsevich, *The Radon Transform and Local Tomography*, CRC Press, Boca Raton, 1996.

[R26] A. G. Ramm, A. B. Smirnova, A numerical method for solving nonlinear ill-posed problems, *Numerical Funct. Anal. and Optimiz.*, 20, N3, (1999), 317-332.

[R27] A. G. Ramm, A numerical method for some nonlinear problems, *Math. Models and Meth. in Appl.Sci.*, 9, N2, (1999), 325-335.

[R28] A. G. Ramm, Inequalities for the derivatives, *Math. Ineq. and Appl.*, 3, N1, (2000), 129-132.

[R29] A. G. Ramm, Linear ill-posed problems and dynamical systems, *Jour. Math. Anal. Appl.*, 258, N1, (2001), 448-456.

[R30] A. G. Ramm, A. B. Smirnova, On stable numerical differentiation, *Math. of Comput.*, 70, (2001), 1131-1153.

[R31] A. G. Ramm, Stability of solutions to inverse scattering problems with fixed-energy data, *Milan Journ of Math.*, 70, (2002), 97-161.

[R32] A. G. Ramm, A. B. Smirnova, Continuous regularized Gauss-Newton-type algorithm for nonlinear ill-posed equations with simultaneous updates of inverse derivative, *Intern. Jour. of Pure and Appl Math.*, 2, N1, (2002), 23-34.

[R33] A. G. Ramm, Regularization of ill-posed problems with unbounded operators, *J. Math. Anal. Appl.*, 271, (2002), 547-550.

[R34] A. G. Ramm, A. Smirnova, Stable numerical differentiation: when is it possible? *Jour. Korean SIAM*, 7, N1, (2003), 47-61.

[R35] A. G. Ramm, On a new notion of regularizer, *J.Phys A*, 36 (2003), 2191-2195.

[R36] A. G. Ramm, On the discrepancy principle, *Nonlinear Functional Anal. and Applic.*, 8, N2, (2003), 307-312.

[R37] A. G. Ramm, Global convergence for ill-posed equations with monotone operators: the dynamical systems method, *J. Phys A*, 36, (2003), L249-L254.

[R38] A. G. Ramm, Dynamical systems method for solving nonlinear operator equations, *International Jour. of Applied Math. Sci.*, 1, N1, (2004), 97-110.

[R39] A. G. Ramm, Dynamical systems method for solving operator equations, *Communic. in Nonlinear Sci. and Numer. Simulation*, 9, N2, (2004), 383-402.

[R40] A. G. Ramm, Inequalities for solutions to some nonlinear equations, *Nonlinear Functional Anal. and Applic.*, 9, N2, (2004), 233-243.

[R41] A. G. Ramm, Continuity of solutions to operator equations with respect to a parameter, *Internat. Jour. of Pure and Appl. Math. Sci.*, 1, N1, (2004), 1-5.

[R42] A. G. Ramm, Dynamical systems method and surjectivity of non-linear maps, *Communic. in Nonlinear Sci. and Numer. Simulation*, 10, N8, (2005),931-934.

[R43] A. G. Ramm, DSM for ill-posed equations with monotone operators, *Comm. in Nonlinear Sci. and Numer. Simulation*, 10, N8, (2005),935-940.

[R44] A. G. Ramm, *Inverse problems*, Springer, New York, 2005.

[R45] A. G. Ramm, Inequalities for the derivatives and stable differentiation of piecewise-smooth discontinuous functions, *Math. Ineq and Applic.*, 8, N1, (2005), 169-172.

[R46] A. G. Ramm, Discrepancy principle for the dynamical systems method, *Communic. in Nonlinear Sci. and Numer. Simulation*, 10, N1, (2005), 95-101.

[R47] A. G. Ramm, *Wave scattering by small bodies of arbitrary shapes*, World Sci. Publishers, Singapore, 2005.

[R48] A. G. Ramm, A new discrepancy principle, *J. Math. Anal. Appl.*, 310, (2005), 342-345.

[R49] A. G. Ramm, Dynamical systems method (DSM) and nonlinear problems, in the book: *Spectral Theory and Nonlinear Analysis*, World Scientific Publishers, Singapore, 2005, 201-228. (ed J. Lopez-Gomez).

[R50] A. G. Ramm, *Random fields estimation*, World Sci. Publishers, Singapore, 2005.

[R51] A. G. Ramm, Uniqueness of the solution to inverse obstacle scattering problem, *Phys. Lett A*, 347, N4-6, (2005), 157-159.

[R52] A. G. Ramm, Dynamical systems method for nonlinear equations in Banach spaces, *Communic. in Nonlinear Sci. and Numer. Simulation*, 11, N3, (2006), 306-310.

[R53] A. G. Ramm, Dynamical systems method and a homeomorphism theorem, *Amer. Math. Monthly*, (2006)

[R54] A. G. Ramm, A nonlinear singular perturbation problem, *Asymptotic Analysis*, 47, N1-2, (2006), 49-53.

[R55] A. G. Ramm, Dynamical systems method (DSM) for unbounded operators, *Proc.Amer. Math. Soc.*, 134, N4, (2006), 1059-1063.

[R56] A. G. Ramm, Existence of a solution to a nonlinear equation, *Jour. Math. Anal. Appl.*, 316, (2006), 764-767.

[R57] A. G. Ramm, Finding discontinuities of piecewise-smooth functions, *JIPAM (Journ. of Inequalities in Pure and Appl. Math.)*

[R58] A. G. Ramm, On unbounded operators and applications, (preprint)

[R59] A. G. Ramm, Ill-posed problems with unbounded operators, *Journ. Math. Anal. Appl.*

[R60] A. G. Ramm, Iterative solution of linear equations with unbounded operators, (preprint).

[R61] A. G. Ramm, Compactness of embeddings, Nonlinear Functional Analysis and Applications, 11, N4, (2006).

[R62] A. G. Ramm, Dynamical systems method for solving linear ill-posed problems, (preprint).

[R63] A. G. Ramm, Dynamical systems method (DSM) for general nonlinear equations, (preprint).

[R64] A. G. Ramm, Discrepancy principle for DSM, http://arxiv.org/abs/math.FA/0603632.

[R65] A. G. Ramm, Distribution of particles which produces a "smart" material, http://arxiv.org/abs/math.AP/0606023

[Ro] E. Rothe, *Introduction to various aspects of degree theory in Banach spaces*, Amer. Math. Soc., Providence, RI, 1986.

[Sh] R. Showalter, *Monotone operators in Banach space and nonlinear partial differential equations*, Amer. Math. Soc., Providence, RI, 1997.

[Sp] E. Spanier, *Algebraic topology*, McGraw-Hill, New York, 1966.

[Te] R. Temam, *Infinite-dimensional dynamical systems in physics and mechanics*, Springer, New York, 1997.

[Tik] V. Tikhomirov, *Some questions in approximation theory*, Nauka, Moscow, 1976.

[TLY] A. Tikhonov, A. Leonov, A. Yagola, *Nonlinear ill-posed problems*, Chapman and Hall, London, 1998.

[Ti] A. Timan, *Theory of approximation of functions of a real variable*, Dover, Mineola 1993.

[T] A. Turetsky, *Interpolation theory in problems*, High School, Minck, 1968. (Russian)

[U] S. Ulm, *On iterative methods with successive approximation of the inverse operator*, Izv. Acad. Nauk Eston SSR, 16, (1967), 403-411.

[V] M. Vainberg, *Variational method and method of monotone operators*, Wiley, New York, 1973.

[VV] G. Vainikko, A. Veretennikov, *Iterative procedures in ill-posed problems*, Nauka, Moscow, 1986. (Russian)

[VA] V. Vasin, A. Ageev, *Ill-posed problems with a priori information*, Nauka, Ekaterinburg, 1993. (Russian)

[Y] K. Yosida, *Functional analysis*, Springer, New York, 1980.

[Z] E. Zeidler, *Nonlinear functional analysis, I-V*, Springer, New York, 1985.

[ZS] Zheng Sonmu, *Nonlinear evolution equations*, Chapman Hall, Boca Raton, 1994.

Index

Mathematics in Science and Engineering
Edited by C.K. Chui, Stanford University

Recent titles:
C. De Coster and P. Habets, *Two-Point Boundary Value Problems: Lower and Upper Solutions*
Wei-Bin Zang, *Discrete Dynamical Systems, Bifurcations and Chaos in Economics*
I. Podlubny, *Fractional Differential Equations*
E. Castillo, A. Iglesias, R. Ruíz-Cobo, *Functional Equations in Applied Sciences*
V. Hutson, J.S. Pym, M.J. Cloud, *Applications of Functional Analysis and Operator Theory (Second Edition)*
V. Lakshmikantham and S.K. Sen, *Computational Error and Complexity in Science and Engineering*
T.A. Burton, V*olterra Integral and Differential Equations (Second Edition)*
E.N. Chukwu, *A Mathematical Treatment of Economic Cooperation and Competition Among Nations: with Nigeria, USA, UK, China and Middle East Examples*
V.V. Ivanov and N. Ivanova, *Mathematical Models of the Cell and Cell Associated Objects*
Z. Zong, *Information-Theoretic Methods for Estimating Complicated Probability Distributions*

www.ingramcontent.com/pod-product-compliance
Ingram Content Group UK Ltd.
Pitfield, Milton Keynes, MK11 3LW, UK
UKHW020243290926
15470UKWH00002B/708

9 780444 527950